T0336974

Microgrid Technology and Engineering Application

Microgrid Technology and Engineering Application

Li Fusheng

Li Ruisheng

Zhou Fengquan

AMSTERDAM • BOSTON • HEIDELBERG • LONDON • NEW YORK • OXFORD
PARIS • SAN DIEGO • SAN FRANCISCO • SINGAPORE • SYDNEY • TOKYO
Academic Press is an Imprint of Elsevier

Academic Press is an imprint of Elsevier
125, London Wall, EC2Y 5AS, UK
525 B Street, Suite 1800, San Diego, CA 92101-4495, USA
225 Wyman Street, Waltham, MA 02451, USA
The Boulevard, Langford Lane, Kidlington, Oxford OX5 1GB, UK

British Library Cataloguing-in-Publication Data
A catalogue record for this book is available from the British Library

Library of Congress Cataloging-in-Publication Data
A catalog record for this book is available from the Library of Congress

ISBN: 978-0-12-803598-6

For information on all Academic Press publications
visit our website at http://store.elsevier.com/

Typeset by Thomson Digital

Printed and bound in the United States

Publisher: Joe Hayton
Acquisitions Editor: Lisa Reading
Editorial Project Manager: Hong Jin
Production Project Manager: Lisa Jones
Designer: Greg Harris

Contents

Foreword

With energy and environment issues becoming increasingly prominent, developing low-carbon economy and promoting ecological progress to achieve sustainability has been widely recognized all over the world, and use of clean, renewable resources has become an important strategy. To coordinate macrogrids and distributed generation (DG), and maximize the strengths of DG for economy, energy, and environment, scholars put forward the concept of "microgrid." It is a single controllable electric power system comprising DG, load, energy storage (ES), and control devices. It can reduce feeder loss, and enhance local power supply reliability and energy efficiency.

With the development of smart grid, microgrid and its key enabling technologies are becoming a heated topic around the world. In an effort to further promote microgrid technologies and provide guidance for construction of microgrids, the authors write this book based on their practices and experience. In this book, the basic concept, key technologies, related standards, and practical design methods and principles of microgrid are comprehensively and systematically discussed, and a typical design case is presented and analyzed.

This book is organized into 11 chapters, as detailed below: Chapter 1 "Overview of Microgrid" describes the history, current status, and trends of microgrids; Chapter 2 "Composition and Classification of the Microgrid" describes the composition, operation and control modes, integration voltage, and classification of microgrids; Chapter 3 "Microgrid and Distributed Generation" introduces types of DG commonly used in microgrids; Chapter 4 "Control and Operation of the Microgrid" introduces control of connection to and disconnection from grids, operation control (three-state control, inverter control), and operation processes in grid-connected mode and islanded mode respectively; Chapter 5 "Protection of the Microgrid" discusses impacts of connection of microgrids on distribution network relay protection, microgrid protection strategies, and configuration scheme of protection for microgrids connected to distribution networks; Chapter 6 "Monitoring and Energy Management of the Microgrid" introduces composition of microgrid monitoring system, energy management, and optimized control methods; Chapter 7 "Communication of the Microgrid" introduces special requirements, design principles, and schemes of microgrid communication system; Chapter 8 "Earthing of a Microgrid" introduces microgrid earthing schemes; Chapter 9 "Harmonic Control of the Microgrid" introduces harmonic control technologies; Chapter 10 "Related Standards and Specifications" presents standards and specifications related to microgrid in and outside China; Chapter 11 "A Practical Case" introduces engineering design and test schemes by taking a specific project as example.

Given the short history of microgrid, many technologies are still under study. This book sums up only existing study results and practices. With studies on this subject going deeper, many new technologies will definitely emerge in the future, and we will amend this book accordingly.

This book is coauthored by Li Fusheng (researcher), Li Ruisheng (professorate senior engineer), and Zhou Fengquan (doctor), and Li Xianwei, Ma Hongwei, Yang Huihong, Tang Yunlong, Zhang Zhiwei, and Wang Huijuan from Smart Grid Research Center of XJ Group contributed significantly to data sorting and diagrams of the book. We also give our special thanks to Dr. Song Xiaowei from Zhongyuan University of Technology who provided a large number of references for us before writing, and Professor Yao Qinglin from Hefei University of Technology who carefully reviewed Chapter 5 "Protection of Microgrid," and gave valuable comments and suggestions. We are also sincerely grateful to the authors of books we referred to while writing this book.

Owing to our limited knowledge and time, there may be some flaws or even errors in this book. Your comments and corrections are highly appreciated.

The Authors

August 2014

Preface

The electric power system is moving into the smart grid era.

Energy efficiency, environmental friendliness, and operational security are the primary concerns for a smart grid. For the sustainability of the human society, it is a must to save nonrenewable energy and reduce emissions of carbon dioxide; to ensure grid security and stability, a proper proportion of power sources must be located in the load center so as to achieve local balance of electricity. In this context, microgrid, encompassing load, power source, and regulation, emerges.

Microgrid is one of the major forms to connect distributed clean energy to grids. Clean energy sources such as PV, wind, and energy storage are integrated into the microgrid on a household, building, or community basis, and the microgrid can centrally manage the distributed resources (DR) in the manner of "virtual power plant (VPP)," so as to adapt to the increasing penetration of distributed clean energy and improve the grids' capability to accommodate these energy forms.

Microgrid is a major means to achieve local balance of electricity. The core of microgrid control and operation can be summed up as "independence and mutual support." "Independence" allows the microgrid to be impervious to grid failure, and operate in islanded mode with little or no load loss, while "mutual support" means the microgrid and main grid can support each other, which not only ensures quality power supply of the microgrid, but also reduces the power transmission burden of the main grid.

Explaining technologies by case and engineering application is one of the features of this book. At present, quite a lot of books on microgrid are available. Instead of following the traditional way to introduce microgrid technologies, this book gives examples on how to analyze the impact of microgrids on traditional distribution networks. Through analysis, problems are raised and technologies are described and explained in solving the problems. Finally, the technologies are demonstrated in engineering application. This is helpful to make clear the viewpoints and provide reference for engineering application.

This book comprises 11 chapters, starting from introduction to roles and significance of microgrids, then classification, integration of DR, operation modes, control and protection, energy management, harmonic control, and earthing modes, then microgrid communication platforms, related standards, and finally analysis of a practical case. With this book, readers can get a clear picture of the concept and technologies of microgrid, and a deep understanding of microgrid engineering. It is a good book for people working on electrical system planning and design, equipment manufacture, and operation management, as well as college students and teachers of this discipline.

I have gained a lot from reading this book and therefore recommend it to you.

Guo Zhizhong
Harbin Institute of Technology
August 2014

Preface

Chapter 1

Overview of microgrid

A microgrid is a single, controllable, independent power system comprising distributed generation (DG), load, energy storage (ES), and control devices, in which DG and ES are directly connected to the user side in parallel. For the macrogrid, the microgrid can be deemed as a controlled cell; and for the user side, the microgrid can meet its unique demands, for example, less feeder loss and higher local reliability. Being capable of autonomous control, protection, and management, a microgrid can operate either in parallel with the main grid or in an intentional islanded mode.

A microgrid can be considered as a small electric power system that incorporates generation, transmission, and distribution, and can achieve power balance and optimal energy allocation over a given area, or as a virtual power source or load in the distribution network. Also, it can consist of one or more virtual power plants (VPPs) to meet the demand of a load center, which can be important offices, factories, or remote residences where the traditional way of electricity supply is expensive. Compared with traditional transmission and distribution (T&D) networks, a microgrid has a much more flexible structure.

1.1 HISTORY

In 2001, Professor R.H. Lasseter of the University of Wisconsin-Madison proposed the concept of the "microgrid." Later, the Consortium for Electric Reliability Technology Solutions (CERTS) and the European Commission Project Micro-Grid also gave their definitions of a microgrid.

In 2002, the National Technical University of Athens (NTUA) built a small laboratory microgrid project known as the NTUA Power System Laboratory Facility for tests on the control of distributed resource (DR) and load with multiagent technology.

In 2003, the University of Wisconsin established a small laboratory microgrid (NREL Laboratory Microgrid) with a capacity of 80 kVA, for tests

Microgrid Technology and Engineering Application

on the control of various types of DRs in different operation modes; another 480 V laboratory microgrid was established in the Walnut test site, Columbus, Ohio, for tests on the dynamic characteristics of various components of a microgrid.

In the same year, multiple demonstration projects were built across the world, including the 7.2 kV microgrid in Mad River Park, Vermont, USA; the 400 V microgrid in Kythnos Islands, Greece; as well as the Aichi, Kyotango, and Hachinohe projects in Japan.

In 2004, the CESI RICERCA test facility was built in Milan, Italy, which can be restructured into different topologies for steady-state and transient-operation tests and power quality analysis.

In 2005, the Imperial College London control and power research center was set up in London, UK, for distribution network prototype tests and load tests.

Over the same period, multiple demonstration projects were successively built all over the world, including Japan's Sendai system (2004), Shimizu Microgrid (2005), and Tokyo Gas Microgrid (2006); Spain's Labein Microgrid (2005); USA's Sandia National Laboratories (2005) and Palmdale's Clearwell Pumping Station (2006); and Germany's Manheim Microgrid (2006).

Since 2006, the microgrid has been successively incorporated into China's 863 Program (State High-Tech Development Plan) and 973 Program (National Basic Research Program). In 2006, Tsinghua University began studies on the microgrid and established a laboratory microgrid encompassing DG, ES, and loads utilizing the facilities in the National Key Laboratory on Power System and Generating Equipment Safety Control and Simulation under the Department of Electrical Engineering.

In 2008, Tianjin University and Hefei University of Technology conducted tests and studies on the microgrid. Tianjin University focused on scientific dispatch of various energy resources in the hope of improving energy efficiency, meeting various demands, and improving reliability, while Hefei University of Technology placed the focus on operation control and energy management.

In 2010, the State Grid Corporation of China (SGCC) built a demonstration project in Zhengzhou for study on operation control of a microgrid combining distributed PV (photovoltaic) generation and energy storage and engineering application and another in Xi'an for study on control technologies for microgrid combining distributed generation/energy storage.

In 2010, the China Southern Power Grid Company built a distributed energy supply – combined cooling and power (CCP) demonstration project in Foshan as a subject under China's 863 Program.

1.2 CURRENT SITUATION OF MICROGRID OUTSIDE CHINA AND ANALYSIS

The world's power sector has been facing great challenges like increasing loads, environmental issues, low energy efficiency, and users' higher requirements on power quality. Microgrids can utilize and control DG in an effective, flexible, and smart manner, and hence, can best address these problems. Many countries are now carrying out studies on the microgrid and their own concepts and goals of a microgrid. As a new technology, the microgrid is showing distinct features in different countries.

1.2.1 USA

The United States is where the concept of "microgrid" originated, and its definition is the most authoritative among all others. The architecture proposed by CERTS consists of power electronic technologies-based micro sources with a capacity of 500 kW or below and loads, and integrates power electronic technologies-based control schemes. Power electronic technologies are indispensable to smart and flexible control and the basis for the "plug and play" and "peer to peer" control and design concepts. CERTS's preliminary study results have been verified with the laboratory microgrid. The Department of Energy (DOE) took microgrid engineering seriously. In 2003, then US President Bush set the goal of grid modernization, that is, to widely integrate IT technologies and communication technologies into power systems to achieve grid smartness. In the later published "Grid 2030," the DOE developed power system study and development plans for the coming decades, in which the microgrid is an important part. On the microgrid meeting convened in 2006, the DOE gave detailed accounts of its microgrid development plans. In view of grid modernization, improving reliability for critical loads, meeting various customized quality demands, minimizing the cost, and realizing smartness will be the focus of the United States' future microgrid.

Figure 1.1 shows the microgrid model proposed by CERTS. This model shows that power electronics interfaces are provided for all micro sources, including PV, wind, small rotary machines, and various types of ESs. The core equipment is a smart static switch that controls the connection to and disconnection from the main grid. For each type of micro source, digital, smart relay protections are used to isolate the protected area from

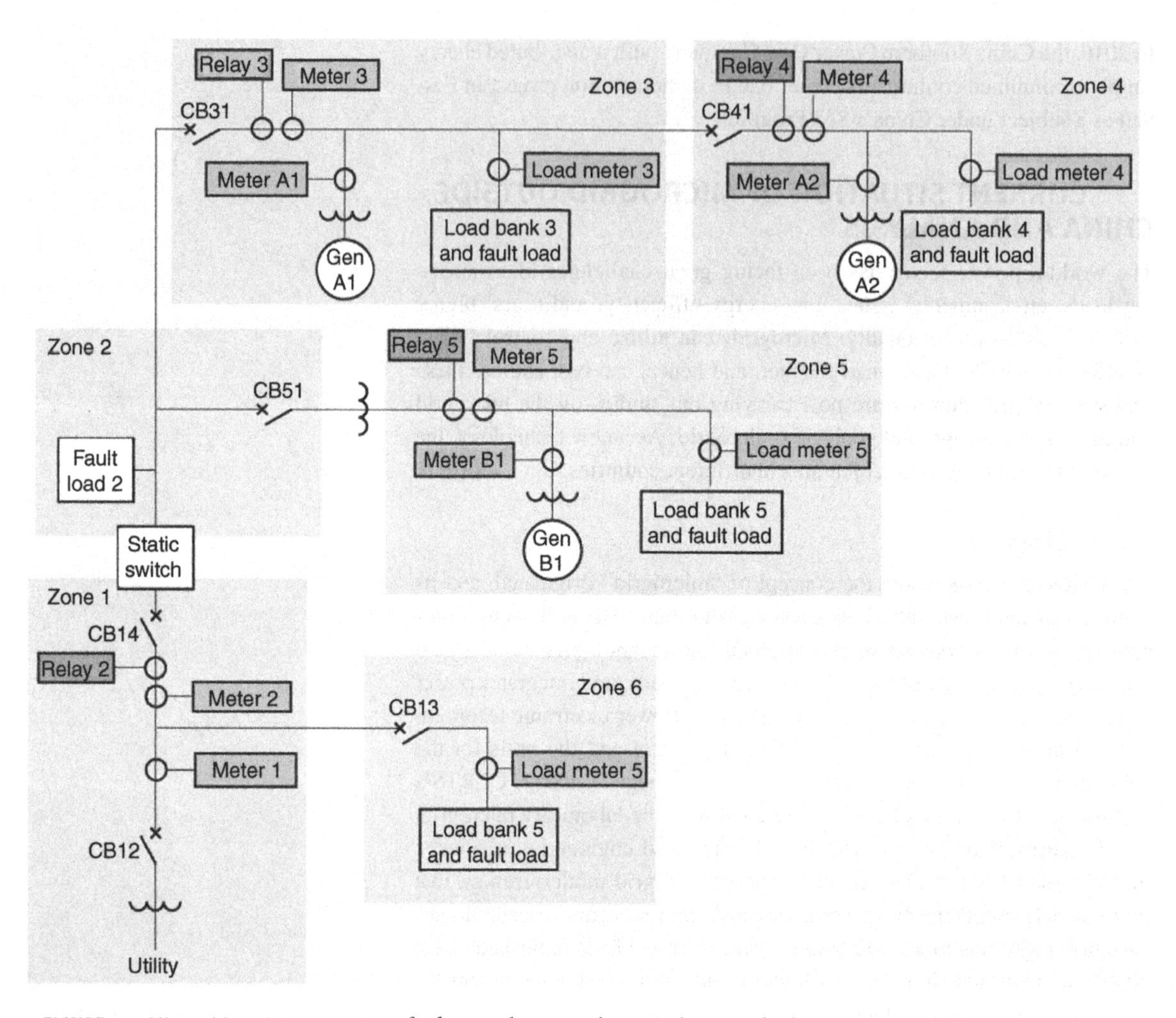

■ **FIGURE 1.1 Microgrid model proposed by CERTS.**

faults, and protection equipment is interconnected via special digital communication links.

1.2.2 Japan

Given the increasing energy shortage and load, Japan studied the microgrid concept with the aim of diversifying energy mix, reducing pollution, and meeting customized demands. In Japan, independent power systems based on traditional sources are also considered as a microgrid, which is a huge extension to the CERTS's definition. On this basis, Japan has implemented multiple microgrid projects. In addition, Japanese scholars put forward the concept of Flexible Reliability and Intelligent Electrical Energy Delivery System (FRIENDS), that is, to add flexible AC transmission systems

(FACTS) to the distribution network to make full use of their advantages in quick and flexible control, optimize the energy mix of the distribution network, and meet varying power quality demands. So far, FRIENDS has become an important form of deployment of microgrids in Japan, and some researchers are considering including the system in combined heat and power systems for better environmental friendliness and higher energy efficiency. Japan has been committed to using new energy for many years. It set up the New Energy & Industrial Technology Development Organization (NEDO) to coordinate studies and use of new energy among universities, companies, and national key laboratories.

1.2.3 European Union

Considering market demands, power supply security, and environmental protection, the European Union (EU) proposed the "Smart Power Networks" program in 2005, and released the strategies in 2006. It called for efficient and close synergy of centralized generation and DG by making full use of distributed energy resource (DER), smart technologies, and advanced power electronic technologies, and called upon all sectors to actively participate in the electricity market and work together to promote the development of grids. Microgrids will be a major part of the European electricity networks thanks to its smartness and diversified energy mix. Currently, theories on operation, control, protection, security, and communications have been established and verified with the laboratory microgrid. The future focus will be more advanced control strategies, standards, and demonstration projects to build the foundation for large-scale integration of DG and transition from the traditional grid to the smart grid. Figure 1.2 shows the microgrid model proposed by the EU with the efforts of ABB, Fraunhofer IWES, and SMA (Germany); ZIV (Spain); The University of Manchester (the UK); EMforce (Holland); and NTUA (Greece).

The microgrid model shown in Figure 1.2 has a more complete structure, where not all micro sources have power electronics interfaces, all protection equipment is digital and smart, and interequipment communication is via controller area network (CAN). Centralized and decentralized monitoring is configured. In centralized monitoring, the central monitoring unit communicates with various switches, gives orders, and sets the switch action range. The monitoring mode is easy and cheap, but has the disadvantage that operation of all switches relies on the central monitoring unit, the failure of which will cause collapse of the entire protection system. A decentralized monitoring system is composed of multiple central monitoring units fulfilling different functions. When one unit fails, the others will automatically take over, thus avoiding system collapse. This mode offers high reliability but calls for more investment.

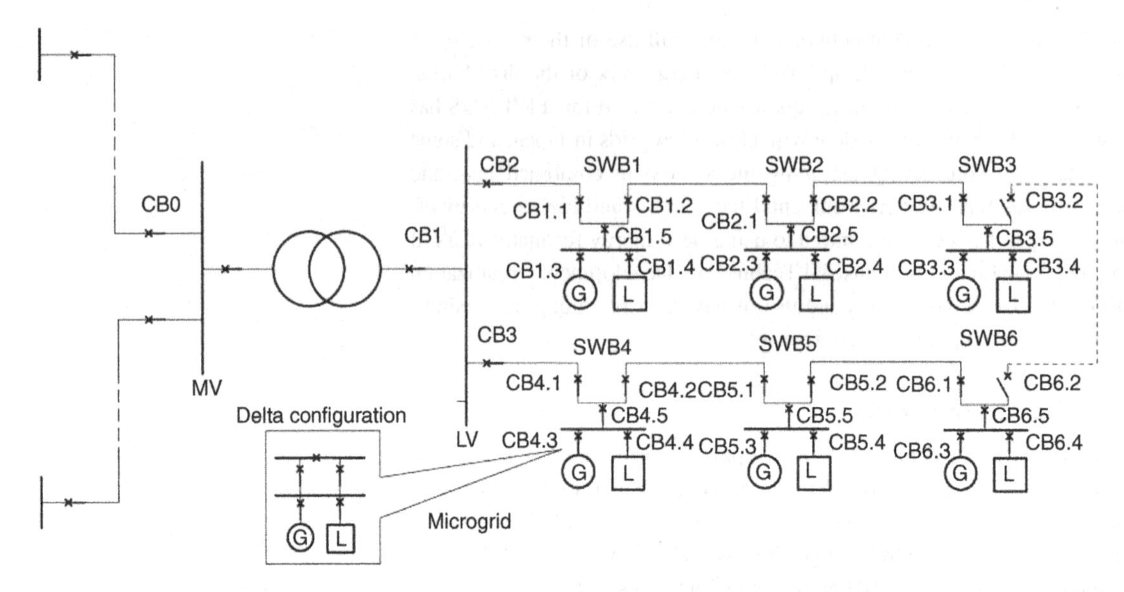

FIGURE 1.2 Microgrid model proposed by the EU. CB, circuit breaker; SWB, switch board; G, micro source; L, load; MV, medium voltage; LV, low voltage.

Apart from the United States, Japan, and the European Union, Canada, Australia, and some other countries have also carried out studies on the microgrid. From their grid strategies, studies on, and practices in microgrid technologies, it can be clearly seen that the development of the microgrid does not represent a revolution to traditional centralized, large-scale grids, but an improvement of the power sector's consciousness of services, energy utilization, and environmental protection. The microgrid is an important means for efficient, environment friendly, and quality power supply by large grids in the future, and hence, a beneficial enhancement to the large grid.

1.3 ANALYSIS OF CURRENT STATUS IN CHINA

China started its studies on the microgrid in 2006, later than other countries. Since this year, the microgrid has been incorporated into the national 863 Program and 973 Program:

863 Program subject in 2006: Distributed power supply system technologies, including technologies and equipment relating to integration, control, and protection of distributed power systems, microgrid technologies.

863 Program subject in 2007: Microgrid technologies, including the following:

1. Structure of microgrids interconnecting multiple energy sources, ESs, and loads and networking technologies;
2. Integration, control, and protection technologies;

3. Control technologies for connection to and disconnection from the grid, and for operation in islanded mode and grid-connected mode;
4. Related power electronics and control technologies;
5. Applicable advanced ESs and control technologies;
6. Internal and external power quality control technologies;
7. Technologies for energy exchange and coordination control between microgrid and macrogrid.

863 Program subject in 2008: distributed energy supply technologies, including the following:

1. Energy matching and control technologies;
2. Microgrid-dependent intermittent power sources and key ES technologies.

973 Program subject in 2009: basic research on distributed power supply system, including the following:

1. Operation characteristics of microgrid and its interaction with macrogrid under high penetration;
2. Theories and methodology on planning distribution systems integrating microgrids;
3. Protection and control for microgrids and distribution systems integrating microgrids;
4. Management methods for comprehensive simulation and energy optimization of distributed power systems.

Currently, many universities, research institutes, and large companies are carrying out studies on the microgrid and have constructed demonstration projects.

In 2006, Tsinghua University began studies on the microgrid and established a laboratory microgrid encompassing DG, ES, and loads utilizing the facilities in the National Key Laboratory on Power System and Generating Equipment Safety Control and Simulation under the Department of Electrical Engineering. Furthermore, Tsinghua, in collaboration with the XJ Group, set up a microgrid simulation platform, established steady-state and dynamic mathematical models of various types of DRs and operation in grid-connected mode, built a simulation environment for microgrids integrating DGs and other power systems and with bidirectional flow, carried out studies on microgrid modeling and operation characteristics analysis, development of simulation platform and operation characteristics analysis, and effects of microgrid on grid load model.

Tianjin University undertook the 973 Program project "Basic research on distributed generation systems," and Huazhong University of Science and

Technology, Xi'an Jiaotong University, and some other organizations participated in the following eight subtopics: (1) interaction with the macrogrid under high penetration; (2) effects of distributed ES on security and stability of the microgrid; (3) optimized planning of distribution systems containing microgrids; (4) theories and technologies of protection for microgrids and distribution systems containing microgrids; (5) microgrid interconnection control and coordinated control of various DRs within the microgrid; (6) power quality analysis and control for microgrids and distribution systems containing microgrids; (7) comprehensive simulation of DG-based microgrids; (8) microgrid economic operation theories and energy optimization management methods.

Hefei University of Technology proposed a new inverter power supply for the microgrid based on the synchronous generator machine-electricity transient model, known as the virtual synchronous generator. It can serve as an uninterrupted power supply for critical loads following a microgrid failure. When the microgrid operates in parallel with the grid, the virtual synchronous generators are power controlled to adjust their output power according to dispatch orders; when the microgrid operates in islanded mode, the inverters are voltage and frequency controlled to provide reference for voltage of the microgrid.

The Institute of Electrical Engineering, Chinese Academy of Sciences, established a 200 kVA laboratory microgrid system, conducted steady-state and dynamic analysis, proposed steady-state and dynamic calculation methods and control and management strategies for microgrid islanded operation, and carried out a lot of studies and tests on control methods of DG and the smooth mode transfer of the microgrid.

Zhejiang University analyzed the active and reactive power circulating current model of parallel-connected inverters in a typical microgrid, and proposed an improved self-regulating droop factor control method considering the problem of instability of output power amplitude and frequency of such inverters adopting traditional droop control strategy.

Sichuan University proposed a multiagent-based uninterrupted substation power coordination system, in which various DRs are capable of automatic compensation for load fluctuation by adopting multiagent coordination strategy, thereby improving load tracing capability and reliability of substations.

In 2009, Zhejiang Electric Power Company developed a laboratory microgrid comprising multiple types of DGs and ESs, which can be flexibly restructured, and therefore can be used to simulate various failures and realize smooth transfer between grid-connected mode and islanded mode. It also carried out tests on various coordination control and protection technologies, quality control, and other advanced applications.

In 2010, Henan Electric Power Company and XJ Group jointly completed SGCC's demonstration project, "Comprehensive study on operation control of microgrid containing PV and ES and engineering application" (as part of the Golden Sun Demonstration Project launched by the Ministry of Finance of the People's Republic of China). This project, located in Henan College of Finance & Taxation, comprises a PV system of 380 kW and an ES system of 2 × 100 kW/100 kWh, and covers the dormitories and canteens in the No. 4 distribution area of the college, including three PV circuits, two ES circuits, and 32 low-voltage (LV) distribution circuits. It communicates with the dispatch center of Zhongmu Electric Power Company.

In the same year, Shaanxi Electric Power Company and the XJ Group jointly completed another demonstration project, "Study on control technologies for microgrid containing distributed generation/energy storage." This project, located at the entrance to Xi'an China International Horticultural Exposition Park, consists of a series of trial projects including smart distribution network, DG, and microgrid, electric vehicle charging station, and experience of intelligent use of electricity, and is intended to exhibit SGCC's new technologies and findings on smart grid to the public. In this project, a 50 kW PV power system is arranged on the shed of the electric vehicle charging station, and six wind turbine generators with an aggregate capacity of 12 kW and a 30 kW/60 kWh energy storage system are arranged around the charging station.

These two demonstration projects prove that the microgrid system can achieve optimal operation in grid-connected mode and stable operation in islanded mode, automatic transfer between the two modes, and control of power exchange with the macrogrid, all being unique to the microgrid. The two demonstration projects have been in commercial operation.

In 2010, CSG's "Distributed energy supply – combined cooling and power demonstration project," a subject under the 863 Program, began commercial operation. This project, located in the Power Supply Bureau of Chancheng District, Foshan, consists of three 200 kW microturbines and one lithium bromide water absorption refrigerator. The system can meet the cooling and power loads of the three buildings in the Bureau, and according to the design objective, the efficiency of primary energy will exceed 75%.

1.4 PROSPECTS

With the advanced IT and communication technologies, electric power systems will develop toward more flexible, cleaner, safer, and more economic smart grids. The smart grid is intended for power systems encompassing generation, transmission, distribution, and consumption, and allows for smart interaction between all links by developing and introducing advanced control technologies, thereby systematically optimizing electricity production, transmission, and consumption. To suit the development of the smart grid, the distribution network has to shift from passive to active, which is favorable to DG and allows for real-time participation of the generation side and user side in optimizing the power system operation. The microgrid is an effective means for an active distribution network, which will help large-scale integration of DG and transition from the traditional grid to smart grid.

The use of various types of DGs and ESs in the microgrid is not only conducive to energy saving and emission reduction, but also significantly motivates the sustainability strategy in China. Compared with traditional centralized power systems, new energy-based DG can largely reduce feeder losses and save investment on T&D networks. It allows for mutual support with the macrogrid, full use of available resources and equipment, and reliable and quality supply, thereby increasing energy efficiency and grid security. In spite of a short history in China, the microgrid technology suits China's needs to develop renewable energy and seek sustainability, and hence, in-depth studies on the microgrid are of great significance.

Chapter 2

Composition and classification of the microgrid

This chapter introduces the composition, structure, operation, and control modes and integration voltages of the microgrid, as well as classification of microgrids by function demand, capacity, and AC/DC type.

2.1 COMPOSITION

A microgrid is composed of distributed generation (DG), loads, energy storage (ES), and control devices. It acts as a single entity with respect to the grid, and connects to the grid via a single point of common coupling (PCC). Figure 2.1 shows the composition and structure of a microgrid.

1. *DG*: It can be various types of new energy, such as photovoltaic (PV), wind, and fuel cell; or combined heat and power (CHP) or combined cooling, heat, and power (CCHP), which provides heat for users locally, thereby increasing efficiency and flexibility of DG.
2. *Loads*: It includes common load and critical load.
3. *ES*: It includes physical, chemical, and electromagnetic forms, for storage of renewable energy, load shifting, and black-start of microgrid.
4. *Control devices*: They constitute the control system for DGs, ESs, and transfer between grid-connected mode and islanded mode, facilitating real-time monitoring and energy management.

2.2 STRUCTURE

Figure 2.2 shows the three-layer microgrid control scheme implemented in the demonstration project based on "multiple microgrid structures and controls." The top layer is the distribution network dispatch layer, which coordinates and dispatches the microgrid to maintain security and economy of the distribution network, and the microgrid is regulated and controlled by the distribution network. The middle layer is the centralized control layer. It forecasts the DG output and load demand, develops operation plans, and adjusts the plans and controls start and stop of DG, load, and ES in real-time

Microgrid Technology and Engineering Application

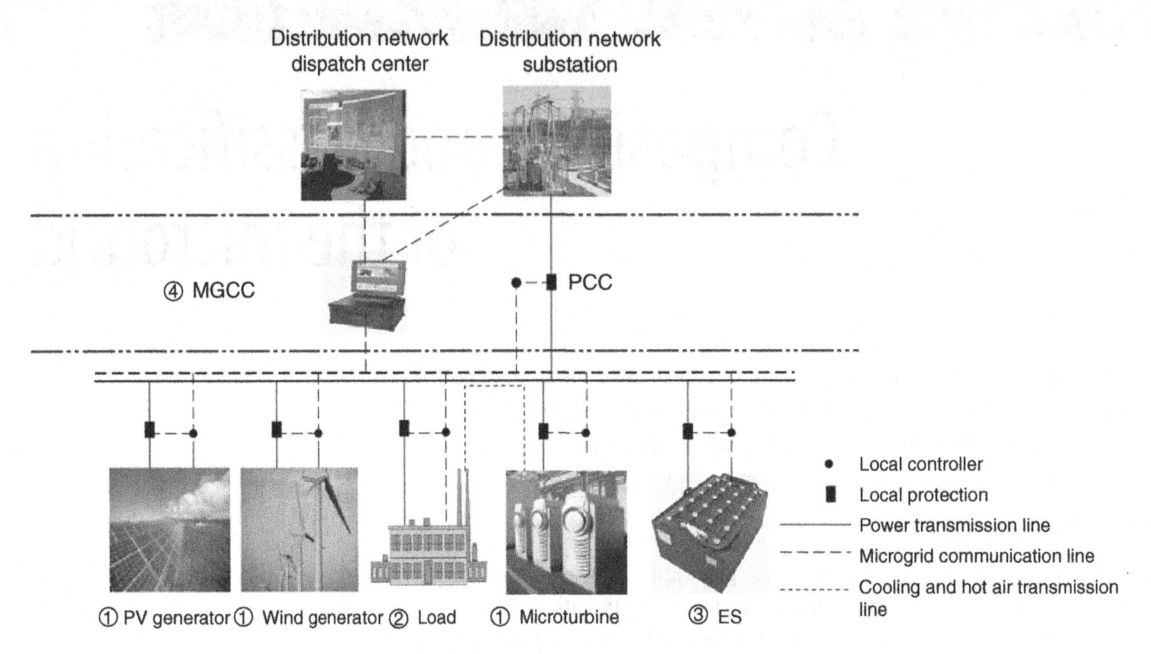

■ **FIGURE 2.1 Composition and structure of microgrid.**

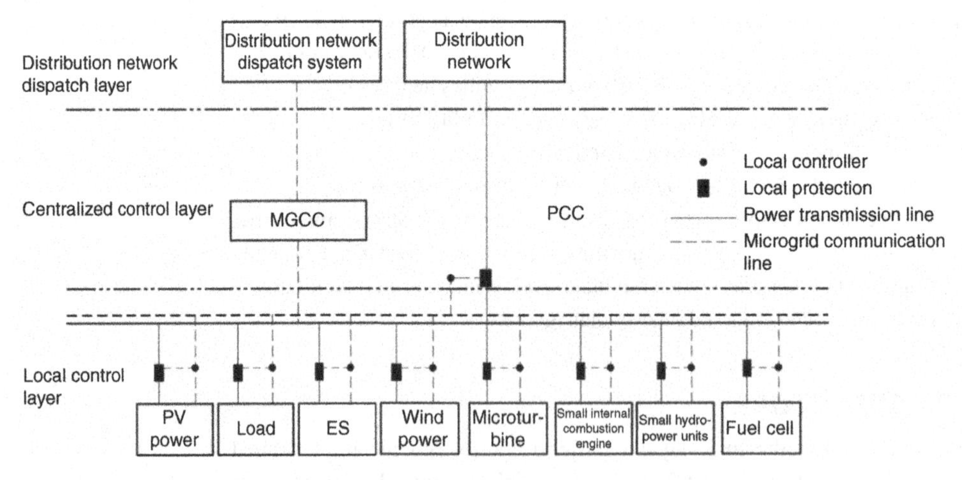

■ **FIGURE 2.2 Structure of three-layer microgrid control.**

based on information such as current, voltage, and power, thereby ensuring voltage and frequency stability of the microgrid. In grid-connected operation, the microgrid operates for best economic efficiency; in islanded operation, the microgrid regulates the output of DGs and consumption of loads to ensure stable and secure operation. The bottom layer, local control layer, executes coordination of DGs, charging and discharging control of ESs, and load control within the microgrid.

2.2.1 Distribution network dispatch layer

The distribution network dispatch layer coordinates and dispatches the microgrid to keep the distribution network in a safe and economic operation, and the microgrid is regulated and controlled by the distribution network.

The microgrid acts as a single controllable and flexible entity with respect to the macrogrid.

1. It can operate either in parallel with the macrogrid or in islanded mode in the case of macrogrid failure or when necessary.
2. In special situations (e.g., an earthquake, snowstorm or flood), the microgrid can serve as reserve to the distribution network, thereby providing effective support to the macrogrid and speeding up recovery of the macrogrid from failure.
3. In case of power shortage in the macrogrid, the microgrid can shift load with its own energy storage, thereby avoiding widespread tripping of the distribution network and reducing the reserve capacity for macrogrid.
4. Under normal circumstances, the layer participates in macrogrid dispatch for best economy.

2.2.2 Centralized control layer

The centralized control layer is the microgrid control center (MGCC) and the core of the microgrid control system. It centrally manages DGs, ESs, and loads, and monitors and controls the entire microgrid. It optimizes the control strategy in real time based on the operating conditions to ensure smooth transfer between grid connection, islanding, and shutdown. In grid-connected operation, the layer regulates the microgrid for best performance; in islanded operation, the layer adjusts the DG output and load consumption to ensure stable and safe operation of the microgrid.

1. In grid-connected operation, the layer dispatches the microgrid for best economic performance and coordinates various DGs and ESs for load shifting to smooth the load curve.
2. During transfer between grid-connected mode and islanded mode, the layer coordinates the local controller to realize quick transfer.
3. In islanded operation, the layer coordinates various DGs, ESs, and loads to maintain supply to important loads and safe operation of the microgrid;
4. When the microgrid stops operation, the layer initiates black-start to rapidly resume operation.

2.2.3 Local control layer

The local control layer of the microgrid is composed of local protection and local controller. The local controller realizes primary regulation of frequency and voltage of DG, while local protection provides quick fault protection for the microgrid. The two work together for quick self-healing of the microgrid from faults. The DG is controlled by the MGCC and adjusts its active and reactive output according to dispatch orders.

1. The local controller controls the automatic transfer between *U/f* control and *P/Q* control.
2. The load controller rejects unimportant loads based on system frequency and voltage to ensure system security.
3. The local control layer communicates with the MGCC through weak ties. The former is responsible for microgrid transient control while the latter for steady-state control and analysis.

2.3 OPERATION MODES

A microgrid may operate either in grid-connected or in islanded mode, and grid-connected operation is further divided into power-matched operation and power-mismatched operation according to power exchange. As shown in Figure 2.3, the microgrid is connected to the distribution network via a PCC, the active and reactive power flowing through the PCC is, respectively, ΔP and ΔQ. When $\Delta P = 0$ and $\Delta Q = 0$, the current through PCC is zero, indicating that the DG output and load reach a balance and no power exchange occurs between the distribution network and microgrid. This is the most economic operation mode of the microgrid, known as power-matched operation. When $\Delta P \neq 0$ or $\Delta Q \neq 0$, the current through the PCC is not zero, indicating that power exchange occurs between the distribution network and the microgrid. This mode is known as power-mismatched operation. In this mode, if $\Delta P < 0$, the excessive active power from DGs after meeting load demand is injected to the distribution network; if $\Delta P > 0$, the electricity from DGs is insufficient for meeting load demand, requiring

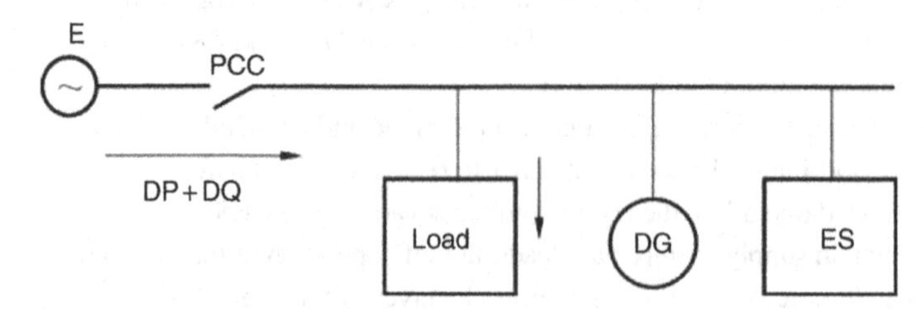

■ **FIGURE 2.3 Power exchange between distribution network and microgrid.**

the distribution network to provide the deficit. Likewise, reactive power is excessive if $\Delta Q < 0$ and deficient if $\Delta Q > 0$. All these operations are power-mismatched operation variants.

2.3.1 Grid-connected operation

In grid-connected mode, the microgrid is connected to and exchanges power with the distribution system of the utility grid via PCC.

Figure 2.4 shows the schematic diagram of transfer between operation modes.

1. When the microgrid stops operation, it can transfer to grid-connected mode directly by grid connection control; when it is connected to the grid, it can be disconnected from the grid by disconnection control.
2. When the microgrid stops operation, it can transfer to islanded mode directly by disconnection control, and when it is in islanded operation, it can be connected to the grid-by-grid connection control.
3. When the microgrid operates in parallel with the grid or in islanded mode, the microgrid can be shut down by shutdown control.

2.3.2 Islanded operation

Islanded operation means that the microgrid is disconnected from the distribution system of the main grid at the PCC following a grid failure or as scheduled, and that the DGs, ESs, and loads within the microgrid operate independently. In islanded mode, since the electricity produced by the microgrid itself is generally small and insufficient to meet the demand of all

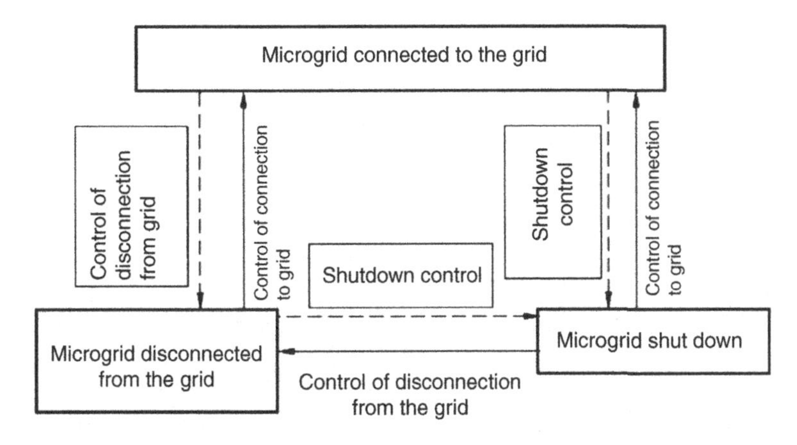

■ FIGURE 2.4 Transfer between operation modes.

loads, it is necessary to prioritize loads based on their importance and ensure uninterrupted supply to important loads.

2.4 CONTROL MODES

2.4.1 Microgrid control modes

Three microgrid control strategies are commonly used, respectively, master–slave mode, peer-to-peer mode, and combined mode. For a small microgrid, master–slave mode is most commonly used.

2.4.1.1 Master–slave mode

In master–slave mode, different DGs within a microgrid are controlled with different methods and assigned with different functions, as shown in Figure 2.5. One or more DGs act as the master while the others as the slave. In grid-connected operation, all DGs are under *P*/*Q* control. In islanded operation, the master DG switches to *U*/*f* control to provide voltage and frequency reference for other DGs. The master DG also traces load fluctuation, and therefore, its power output has to be controllable to some extent, and the DG should be able to respond fast enough to load fluctuation. The slave DGs remain under *P*/*Q* control.

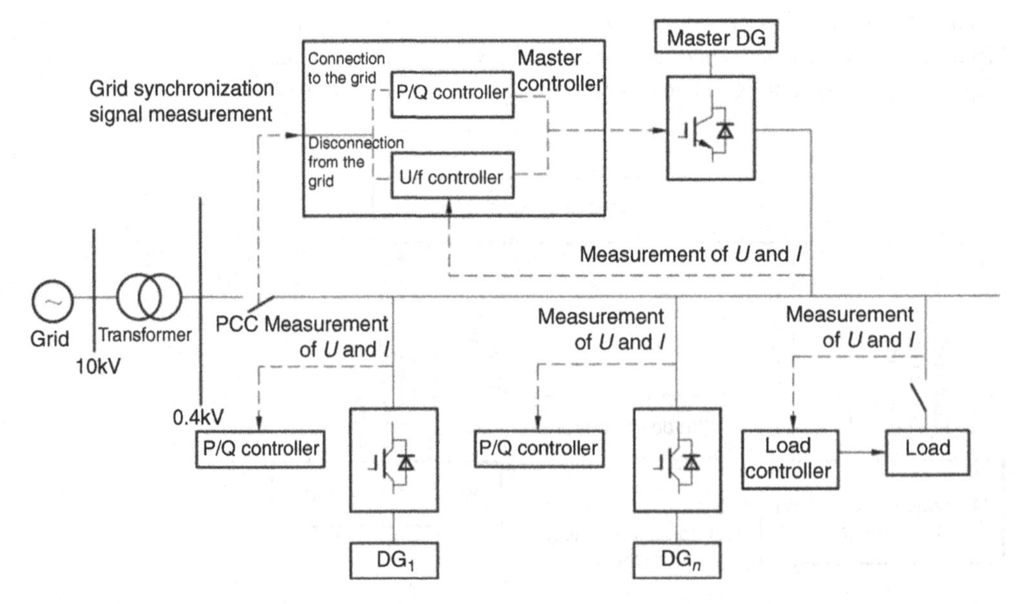

■ **FIGURE 2.5 Architecture of microgrid under master–slave control.**

This control mode has some disadvantages. First, with the master DG under *U/f* control, its voltage output is constant. To increase the power output, the only way is to increase the current output. Instantaneous load fluctuations are usually first balanced by the master DG, and therefore, it has to have a certain adjustable capacity. Second, as the system relies on the master DG to coordinate and control all slave DGs, once the master DG fails, the whole microgrid will collapse. Third, master–slave control requires accurate and timely islanding detection, while islanding detection itself is accompanied by error and time delay. Hence, without a communication channel, transfer between control strategies is likely to fail.

To enable quick transfer between control modes, the following forms of DG could be used as master DGs:

1. DGs of randomness such as PV and wind;
2. ES, microturbine, fuel cell, and other stable and easily controlled DGs; and
3. DG + ES, for example, PV + ES or PV + fuel cell.

The third form is advantageous over the first two in that it can make best use of quick charge and discharge functions of ES and the capability of DGs in maintaining islanded operation of the microgrid for an extended time. In this mode, the ES can provide power support for the microgrid immediately after it switches to islanded operation, thus effectively suppressing significant fluctuation of voltage and frequency resulting from slow dynamic response of DG.

2.4.1.2 Peer-to-peer mode

Peer-to-peer mode is a control strategy based on ideas of "plug-and-play" and "peer-to-peer" used in power electronic technologies. In this mode, all DGs in the microgrid are equal and there is no master or slave DG. All DGs participate in regulation of active power and reactive power in a preset control mode to maintain stability of the system voltage and frequency. Droop control is adopted in the peer-to-peer mode, with the architecture shown in Figure 2.6. In this mode, all DGs under droop control participate in voltage and frequency regulation of the microgrid in islanded operation. When the load changes, the changes will be automatically distributed among the DGs according to the droop factor, that is, all DGs will adjust the frequency and amplitude of their output voltage to establish a new steady state for the microgrid and finally achieve reasonable distribution of output power. The droop control model enables automatic distribution of load variations among DGs, but the voltage and frequency of the system also vary after load variation, and therefore, this control mode is actually a proportional control. The droop control model of DGs can remain unchanged for

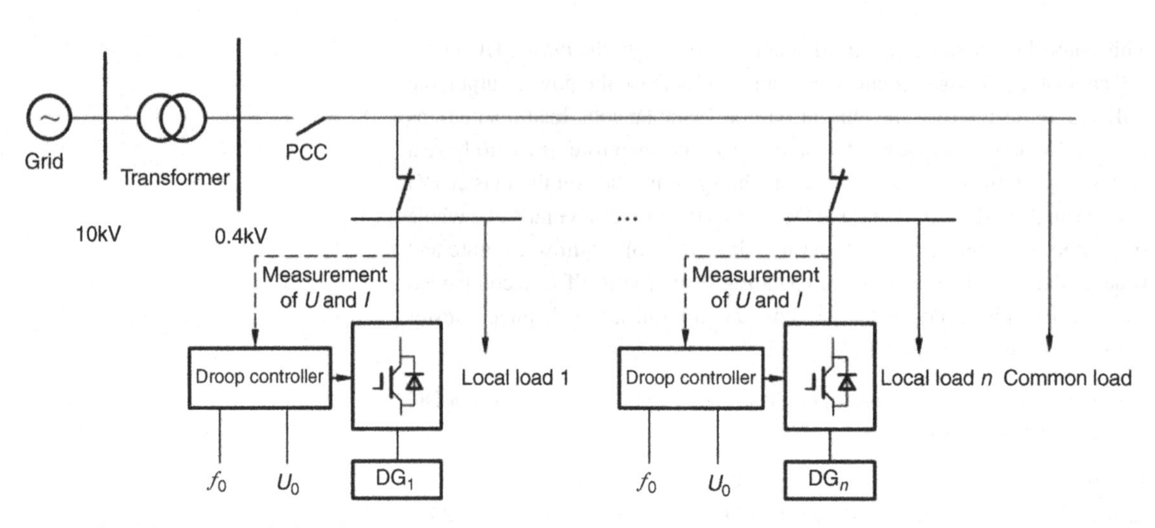

FIGURE 2.6 Architecture of microgrid under peer-to-peer control.

grid-connected operation and islanded operation, making it easy for smooth transfer between the two modes.

The droop control model allows for independent control of DGs according to the voltage and frequency at the PCC, thus making it possible for automatic regulation of voltage and frequency, plug-and-play of DG without communication links, and flexible and convenient deployment of the microgrid. Unlike the master–slave mode where power imbalance is compensated by the master DG, power imbalance is dynamically distributed to all DGs in the peer-to-peer mode. This kind of control is simple, reliable, and easy to deploy, but at the sacrifice of voltage and frequency stability; it is currently under laboratory test.

2.4.1.3 *Combined mode*

Master–slave control and peer-to-peer control have advantages and disadvantages. A microgrid may contain multiple types of DGs, such as DG of randomness (e.g., PV and wind), or stable and easily controlled DG or ES (e.g., microturbine and fuel cell). Control characteristics differ greatly for different types of DG. Apparently, a single control mode cannot meet the operation requirements of a microgrid. In view of the dispersive DGs and loads within a microgrid, different control strategies may be adopted for different types of DGs, that is, master–slave control and peer-to-peer control could be used in conjunction in a microgrid.

2.4.2 Inverter control modes

The DGs integrated to a microgrid may operate either in parallel with the grid or in islanded mode. In the former case, the DGs only need to control their own power output to maintain balance within the microgrid. As the total capacity of a microgrid is much smaller than that of a grid, the rated voltage and frequency are supported and regulated by the grid, and the inverters are usually under *P*/*Q* control. In the latter case, the microgrid is isolated from the grid. To maintain the rated voltage and frequency within the microgrid, one or more DGs need to play the role of the grid to provide rated voltage and frequency. These DGs are usually under *U*/*f* and droop control.

2.4.2.1 *P*/*Q* control

As the interface between the microgrid and macrogrid, the basic function of inverters is to control the active and reactive output. In *P*/*Q* control, the inverters can produce active power and reactive power, and the determination of reference power is the prerequisite for power control. For purpose of power control, the DGs with a mediate or small capacity can be integrated to the grid with a constant power, the grid provides rigid support for voltage and frequency, and the DGs do not participate in frequency and voltage regulation and just inject or absorb power. This can avoid direct participation of DG in the regulation of feeder voltage, thus eliminating adverse impacts on the electric power system.

P/*Q* control is based on the grid voltage oriented *P*/*Q* decoupled control strategy, in which the outer loop adopts power control and the inner loop adopts current control. The mathematical model is like this: the three-phase voltage is first rotated to the dq coordinate through Park transformation to get the following inverter voltage equation:

$$\left.\begin{aligned} v_d &= Ri_d + L\frac{di_d}{dt} - \omega Li_q + u_d \\ v_q &= Ri_q + L\frac{di_q}{dt} + \omega Li_q + u_q \end{aligned}\right\} \tag{2.1}$$

where u_d and u_q are the voltage at the inverter terminal, and ωLi_q and ωLi_d are cross-coupling terms. They will be eliminated by feedforward compensation in subsequent control.

The PI controller is usually used for outer-loop power control. Its mathematical model is expressed as follows:

$$\left.\begin{aligned} i_{dref} &= (P_{ref} - P)\left(k_p + \frac{k_i}{s}\right) \\ i_{qref} &= (Q_{ref} - Q)\left(k_p + \frac{k_i}{s}\right) \end{aligned}\right\} \tag{2.2}$$

where P_{ref} and Q_{ref} are the reference active power and reference reactive power, respectively, and i_{dref} and i_{qref} are the d-axis reference current and q-axis reference current, respectively.

If the grid voltage u is constant, the active output of the inverter is proportional to d-axis current i_d and the reactive output proportional to q-axis current i_q, respectively.

The transfer function between v_{d1}/v_{q1} and i_d/i_q is a first-order lag, which means that the d-axis and q-axis voltages can be controlled by the d-axis and q-axis currents. On this basis, the inner-loop current controller, usually PI controller, can be designed. Its mathematic model is expressed as follows:

$$\left.\begin{aligned} v_{dl} &= (i_{dref} - i_d)\left(k_p + \frac{k_i}{s}\right) \\ v_{ql} &= (i_{qref} - i_q)\left(k_p + \frac{k_i}{s}\right) \end{aligned}\right\} \tag{2.3}$$

Then, by adding compensation terms, the effects of grid voltage and d–q cross-coupling can be eliminated and current decoupling control can be achieved. The inverter control wave can be obtained by reverse Park transformation of d-axis and q-axis voltages, and then the three-phase voltage output of the inverter can be derived by sinusoidal pulse width modulation. Figure 2.7 shows the schematic diagram of P/Q control.

2.4.2.2 *U/f* control

In *U/f* control, the inverters output constant voltage and frequency to ensure continual operation of slave DGs and sensitive loads after the microgrid is isolated from the grid. Given the limited capacity of the microgrid in

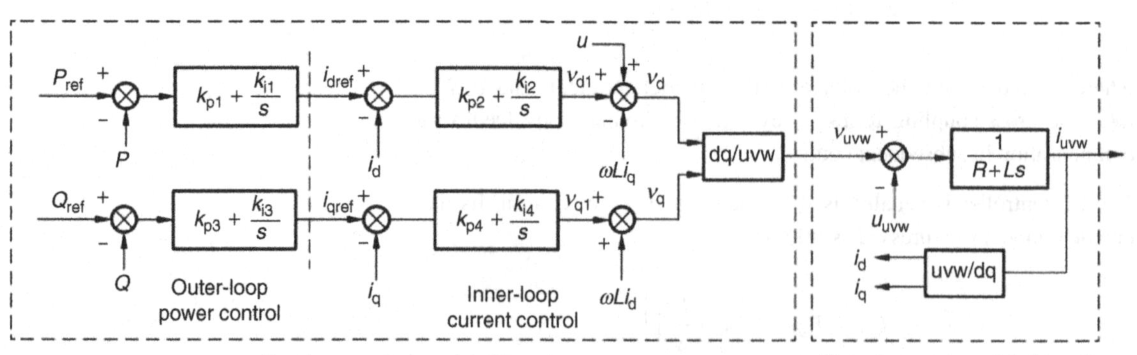

FIGURE 2.7 Schematic diagram of *P/Q* control.

islanded operation, once power shortfall occurs, it is necessary to shed some less important loads to ensure continuous supply to sensitive loads. As such, this control mode requires the ability to respond to and trace load switching.

In this control mode, the AC-side voltage is regulated according to voltage feedback from the inverter to maintain a constant output, and the dual-loop control scheme with outer-loop voltage control and inner-loop current control is often adopted. Outer-loop voltage control can maintain stable voltage output, and inner-loop current control constitutes the current servomechanism system, and can significantly accelerate the dynamic process to defend against disturbances. This dual-loop control can make the best use of system status information, and has a high dynamic performance and steady-state precision. Furthermore, inner-loop current control increases the bandwidth of the inverter control system, thereby speeding up the dynamic response of the inverter, enhancing the inverter's adaptability to nonlinear load disturbance, and reducing harmonic distortion of the output voltage.

The *U/f* control is similar to *P/Q* control in terms of decoupling and control mechanism. Figure 2.8 shows the schematic diagram of *U/f* control, where outer-loop voltage control and inner-loop current control are adopted and the reference voltages U_{ldd}^* and U_{ldq}^* and measured voltages U_{ldd} and U_{ldq} are specified.

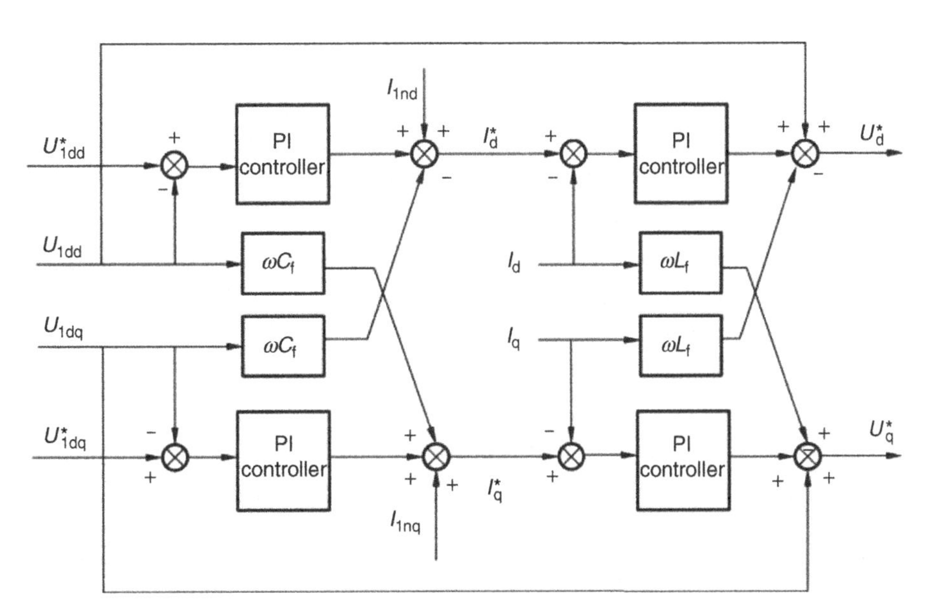

■ **FIGURE 2.8 Schematic diagram of *U/f* control.**

2.4.2.3 Droop control

Droop control is realized by simulating the droop characteristic of generators in a traditional grid and controlling the output voltage and frequency of the voltage source inverter (VSI) according to variation of the output power. The control strategy is based on inverter parallel-connection technology. As all DGs are integrated to the microgrid via inverters, the microgrid in islanded operation is equivalent to multiple inverters being connected in parallel, and the active and reactive output of individual inverters are, respectively

$$\left.\begin{aligned} P_n &= \frac{UU_n}{X_n}\delta_n \\ Q_n &= \frac{UU_n - U^2}{X_n} \end{aligned}\right\} \tag{2.4}$$

where U is the integration voltage, U_n the output voltage of the inverter power supply, X_n the output impedance of the inverter power supply, and δ_n the included angle between U_n and U.

According to Eq. (2.4), the delivery of active power mainly depends on δ_n and that of reactive power mainly depends on the output voltage amplitude of the inverter power supply U_n. U_n can be directly controlled, and the phase can be controlled by adjusting the output angular frequency or frequency of the inverter, as expressed in Eq. (2.5):

$$f_n = \frac{\omega_n}{2\pi} = \frac{d\delta_n}{dt} \tag{2.5}$$

It can be seen from Eqs (2.4) and (2.5) that the output voltage of the inverter can be regulated by regulating its reactive output, and the output frequency can be regulated by regulating its active output. Figure 2.9 shows the droop control characteristic so derived.

Reverse droop control is to control the active and reactive outputs by measuring grid voltage amplitude and frequency to trace the predefined droop

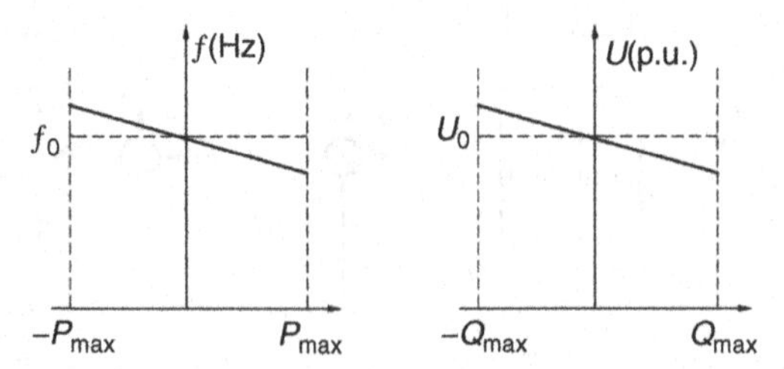

■ FIGURE 2.9 Droop control characteristic.

characteristic. This is a total reversion of the control mode where the output voltage is regulated by measuring the output power and therefore called reverse droop control. As the name implies, the reactive output and active output of the inverter are regulated by regulating the output voltage amplitude and output frequency, respectively.

To make the microgrid operational, inverters may adopt *P/Q* control, droop control or reverse droop control, and with these control modes, the output power of DG can be controlled by simply measuring local data.

2.5 INTEGRATION VOLTAGE CLASS

Microgrids can be integrated into grids at the following three voltages:

1. 380 V (mains)
2. 10 kV
3. A hybrid of 380 V and 10 kV.

Figure 2.10 shows the integration voltages of a microgrid. Figure 2.10a represents integration to a 380 V (mains) LV distribution network, Figure 2.10b represents integration to a 10 kV distribution network with the voltage being increased to 10 kV from 380 V by a step-up transformer, and Figure 2.10c represents integration to a 380 V (mains) LV distribution network and a 10 kV distribution network.

2.6 CLASSIFICATION

Different types of microgrids should be established according to capacity, location, and types of DRs to suit the local situation. Microgrids can be classified as follows.

2.6.1 By function demand

Microgrids are classified into simple microgrid, multi-DG microgrid, and utility microgrid by function demand.

2.6.1.1 *Simple microgrid*

A simple microgrid contains only one type of DG, has simple functions and design, and is intended for use of CCHP or continuous supply to critical loads.

2.6.1.2 *Multi-DG microgrid*

A multi-DG microgrid is composed of multiple simple microgrids or multiple types of complementary, coordinated DGs. Compared with a simple microgrid, the design and operation of such a grid are much more

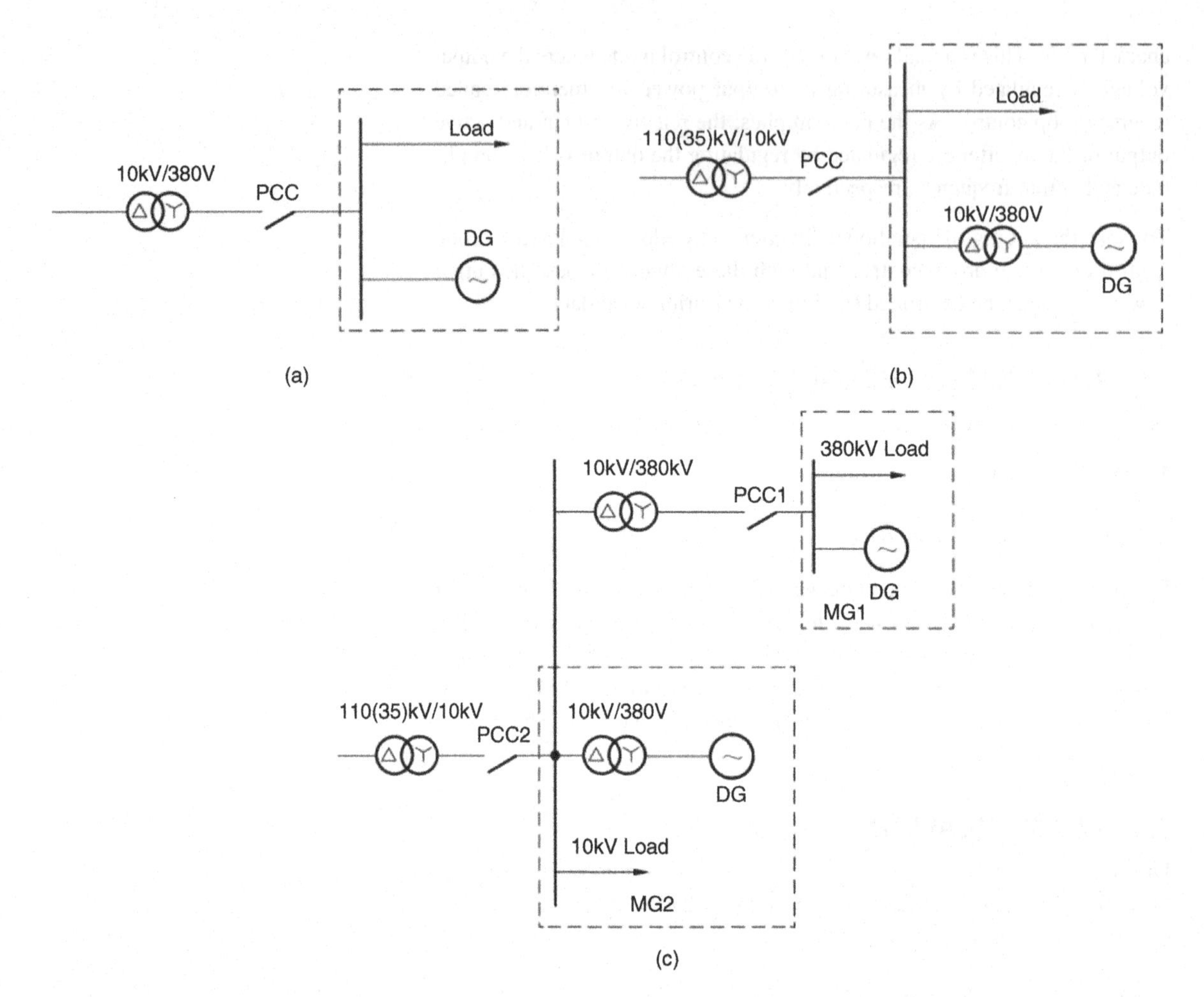

■ FIGURE 2.10 Microgrid integration voltage classes. (a) 380 V, (b) 10 kV, and (c) hybrid of 380 V and 10 kV.

complicated. Some loads need to be identified as sheddable loads in case of emergency to maintain power balance during islanded operation.

2.6.1.3 *Utility microgrid*

All DGs and microgrids that meet specific technical conditions can be integrated into a utility microgrid. In such a microgrid, loads are prioritized based on users' requirements on reliability, and high-priority loads will be powered preferentially in an emergency.

Classifying microgrids by function demand clearly defines the ownership of a microgrid during operation: simple microgrids can be operated and managed by customers, utility microgrids can be operated by utilities, and multi-DG microgrids can be operated either by utilities or customers.

Table 2.1 Classification of Microgrids by Capacity

Type	Capacity (MW)	Grid to be Connected
Simple microgrid	<2	Common grid
Corporate microgrid	2–5	
Feeder area microgrid	5–20	
Substation area microgrid	>20	
Independent microgrid	Depending on loads on an island, a mountainous area or a village	Diesel-fueled grid

2.6.2 By capacity

Microgrids are classified into simple microgrid, corporate microgrid, feeder area microgrid, substation area microgrid, and independent microgrid by capacity, as shown in Table 2.1.

2.6.2.1 *Simple microgrid*

A simple microgrid has a capacity below 2 MW and is intended for independent facilities and institutes with multiple types of loads and of a small area, such as a hospital or school.

2.6.2.2 *Corporate microgrid*

A corporate microgrid has a capacity of 2–5 MW, and comprises CCHPs of varying sizes and some small household loads, generally no commercial or industrial loads.

2.6.2.3 *Feeder area microgrid*

A feeder area microgrid has a capacity of 5–20 MW and comprises CCHPs of varying sizes and some large commercial and industrial loads.

2.6.2.4 *Substation area microgrid*

A substation area microgrid has a capacity above 20 MW and generally comprises common CCHPs and all nearby loads (including household, commercial, and industrial loads).

These four types of microgrids are connected to common grids, and therefore, are collectively called grid-connected microgrid.

2.6.2.5 *Independent microgrid*

An independent microgrid is mainly intended for remote off-grid areas such as an island, a mountainous area, or a village, and the distribution system of the main grid uses a diesel generator or other small units to meet the power demand of such areas.

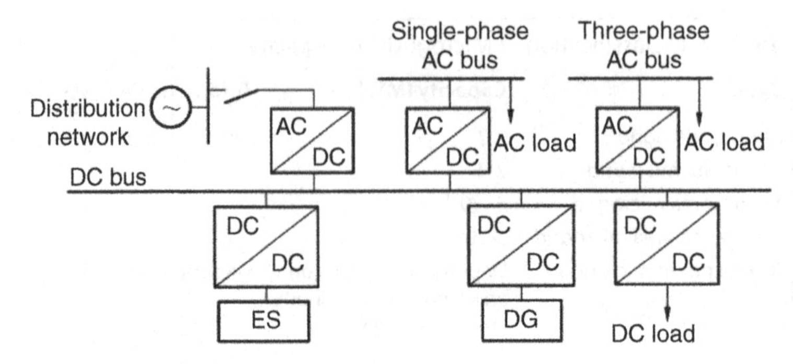

■ FIGURE 2.11 Structure of a DC microgrid.

2.6.3 By AC/DC type

Microgrids are classified into DC microgrid, AC microgrid, and AC/DC hybrid microgrid.

2.6.3.1 DC microgrid

As shown in Figure 2.11, in a DC microgrid, DG, ES, and DC load are connected to a DC bus via a converter and the DC bus is connected to AC loads via an inverter to power both DC and AC loads.

The advantages of the DC microgrid are as follows:

1. As DG control solely depends on DC voltage, it is easier to realize coordinated operation of the DGs.
2. DG and load fluctuations are compensated by ES on the DC side.
3. Compared with an AC microgrid, a DC microgrid is easier to control, does not involve synchronization among DGs, and thus is easier to suppress circulating current.

The disadvantage of the DC microgrid is that inverters are required for the power supply to AC loads.

2.6.3.2 AC microgrid

An AC microgrid connects to the distribution network via an AC bus, and the AC bus controls the microgrid's connection to and disconnection from the distribution network through the circuit breaker at the PCC. Figure 2.12 shows the structure of an AC microgrid, in which DG and ES are connected to the AC bus via inverter. The AC microgrid is a dominant type of microgrid, and the major topic of this book in the following chapters.

The advantage of the AC microgrid is that as the microgrid is connected to the grid through an AC bus, no inverter is required for power supply to AC loads. The disadvantage is that control and operation are difficult.

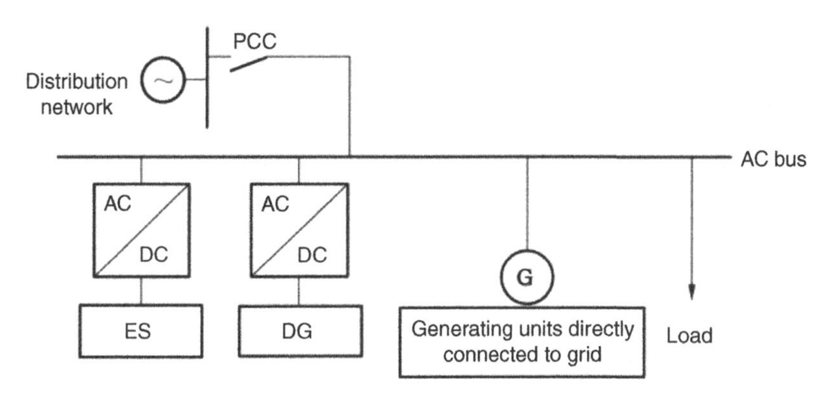

■ **FIGURE 2.12 Structure of an AC microgrid.**

2.6.3.3 *AC/DC hybrid microgrid*

An AC/DC hybrid microgrid is a microgrid consisting of an AC bus and a DC bus. Figure 2.13 shows the structure of such a microgrid, in which the AC bus and DC bus allow for direct supply to AC loads and DC loads. With special power sources connected to an AC bus, an AC/DC hybrid microgrid is essentially a special AC microgrid on the whole.

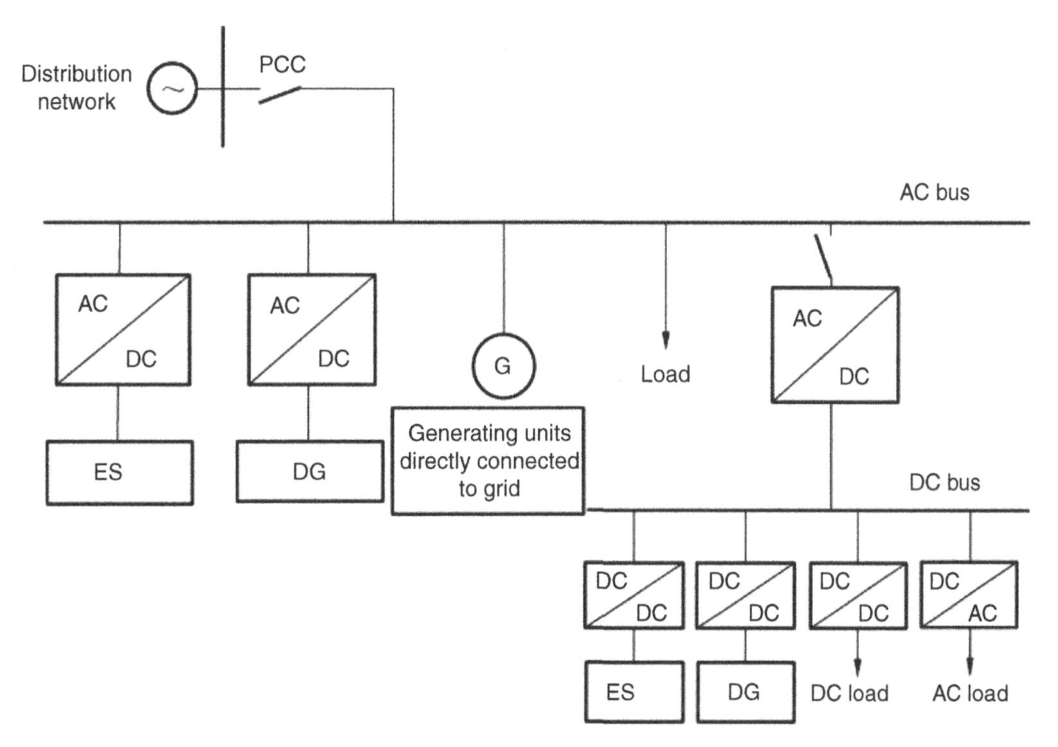

■ **FIGURE 2.13 Structure of an AC/DC hybrid microgrid.**

Chapter 3

Microgrid and distributed generation

3.1 CONCEPTS AND CHARACTERISTICS

Distributed generation (DG) refers to any small electric power system independent of traditional utility grids, which is located on the user side to meet end-users' unique demands. This includes internal combustion engine, microturbine, fuel cell, small hydropower system, photovoltaic (PV) generation, wind generation, waste generation, and biomass generation.

Distributed resource (DR) refers to a combined DG and energy storage (ES) system, that is, DR = DG + ES. It includes all DG technologies and can store energy in a battery, flywheel, regenerative fuel cell, superconducting magnetic storage device, and other devices.

Distributed energy resources (DER) means generation of electricity or heat on the user side for local use. It includes all DG and DR technologies, and systems connected to a utility grid with which users can sell surplus power to utilities. From these definitions, it can be inferred that DG is a subset of DR, which is then a subset of DER.

No definition has been given to DER in China. All of its definitions are introduced from other countries. With a growing interest in DER, studies on this subject are going deeper. The following sums up the characteristics of a distributed energy system:

1. Comprehensive and efficient energy use. A traditional centralized power system can only provide power, while other types of energy, such as heat, in particular cooling and hot water, when required, can only be provided by power, making it impossible for comprehensive and cascaded energy use. While a distributed energy system, with a small size and high flexibility, can not only meet load demand, but also solve the difficulty of long-distance transmission of cooling or heating sources. A large power plant generally has a generation efficiency ranging from 35% to 55%, while the end-use efficiency is only 30–47% after deducting feeder loss. The efficiency of

Microgrid Technology and Engineering Application

a distributed energy system can reach above 80% without any transmission loss.

2. An improvement to grid security and stability. Several widespread blackouts have occurred in the world in recent years; for example, the blackout in east California, USA, revealing the weakness of modern interconnected power systems. In addition, after the September 11 attacks, power supply became a national security issue and drew great attention in all countries, and the rapid expansion of the grid poses a threat to grid security and stability. Deploying a distributed energy system on the user side as a supplement to the macrogrid can significantly enhance reliability and continuity of power supply to critical loads in the event of grid collapse or disasters such as an earthquake, snowstorm, sabotage, or war.
3. Small capacity, small area, low initial investment, no long-distance transmission loss and investment on transmission and distribution (T&D) network, and ability to meet special demands. A distributed energy system is located as close as possible to loads for best coordination with users. Unlike a centralized energy system, a distributed energy system obviates the need for long-distance transmission and distribution, thus causing no feeder loss, requiring no investment on T&D network, and contributing to good economy and flexible, energy-efficient, and comprehensive services for end users. It is an ideal choice for remote western areas where it is infeasible to erect a grid. Another benefit is that the waste energy, such as residual heat, residual pressure, and combustible waste gas, can be reused.
4. Environmental friendliness, diversified energy mix, a new way to utilize renewable energy. Using clean fuels as the energy, a distributed energy system is environmentally friendly. Compared with fossil fuels, renewable energy forms, such as PV, geothermal energy, and wind, have a smaller energy density and widespread distribution. However, the current renewable energy systems have a small capacity and low efficiency, and thus, are not suitable for centralized power supply. In contrast, a distributed energy system has a small capacity and is suitable for integration of renewable energy.

DG has many advantages, but at the same time is hard to control and fluctuates randomly. Therefore, a higher penetration of DG may jeopardize grid stability. For the grid, DG is an uncontrollable power source. At present, DG is often limited or isolated to reduce its impacts on the grid. As stipulated in IEEE 1547 *IEEE Standard for Interconnecting Distributed Resources with Electric Power Systems*, when the electric power system fails, the DG must be taken out of service. A microgrid controls DG, ES, and loads coordinately with the control system to form a single controllable power source and is directly arranged on the user side. For the grid, the microgrid is a controllable entity; and for the user side, the microgrid can meet its unique demands, reduce feeder loss, and ensure local voltage stability.

A microgrid has the following advantages:

1. ES and DG are combined, thus addressing the problem of significant fluctuations of DG outputs.
2. DG is connected to the grid through power electronics, which can flexibly control the active and reactive outputs and voltage output of DG, improving grid reliability.
3. Small combined heat and power (CHP) plants are generally located in heat load centers, for example, for air conditioning and power supply in a commercial building. This way, electricity and heat can be fully utilized.
4. In case of grid failure or a disaster, the microgrid can operate in islanded mode, ensuring power supply reliability.

Various types of DGs are discussed later.

3.2 PHOTOVOLTAICS

PV is a means of electricity generation by directly converting solar energy to electricity. The solar cell is the core component for light-to-electricity conversion. Currently, crystalline silicon solar cell is the dominant type in the market, and other types include amorphous silicon thin film solar cell and compound thin film PV cell. A PV power system may operate independently or in parallel with the grid.

3.2.1 Independent PV power system

An independent PV power system is not connected to a traditional electric power system and is mostly deployed in remote off-grid areas to meet local demands. PV electrification is possible only during the daytime, while power is required around the clock. To solve this issue, independent PV systems must be provided with ES. Figure 3.1 shows the structure of an independent PV power system, mainly consisting of solar cell array, DC combiner box, controller, battery, off-grid inverter, and AC distribution box. The PV components produce direct current to charge the battery and, after DC to AC conversion by the inverter, serve the AC loads.

Solar cell array: Consists of two or more solar cell modules formed by encapsulating solar cells. At present, single crystalline or polycrystalline silicon solar cells are used, which are made of waterproof glass on

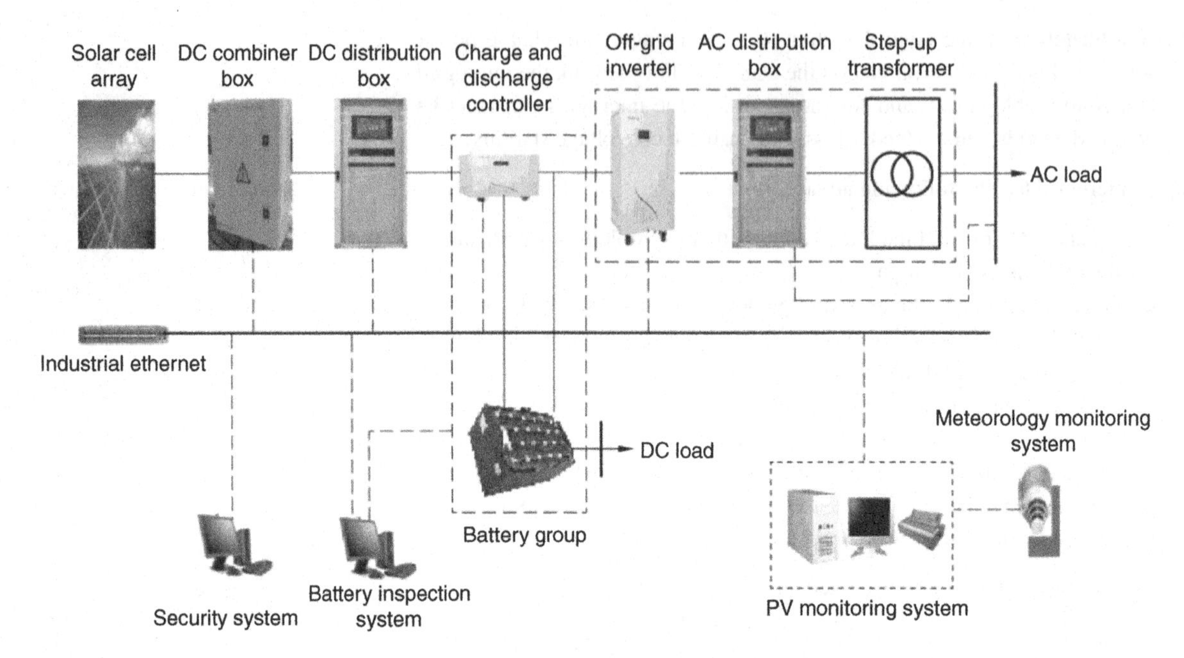

■ **FIGURE 3.1 Structure of independent PV power system.**

the front contact and soft material on the back contact. It is the most fundamental component of a PV power system for conversion of solar energy to electricity.

DC combiner box: Combines multiple circuits of low-current DC outputs of solar cell array into one or more circuits of high-current outputs. Its output may then be collected to the next-level combiner box or the inverter, and it can protect against and monitor the occurrence of overcurrent, countercurrent, and lightning strike.

Controller: Controls the charge and discharge voltage and current of the battery, balances the energy of the system, collects system status information, and controls, protects, and monitors the charge and discharge processes of ESs.

Battery: Stores the intermittent and uncertain energy produced by solar cells to ensure power supply balance and continuity.

Off-grid inverter: Inverts direct current to alternating current to serve AC loads.

AC distribution box: An enclosed or semienclosed metal box housing AC-side switchgear, meters, protections, and other auxiliary equipment for ease of maintenance and management.

3.2.2 Grid-connected PV power system

A grid-connected PV power system is connected to the grid and injects electricity to the grid. It is the mainstream of PV power systems. Grid-connected

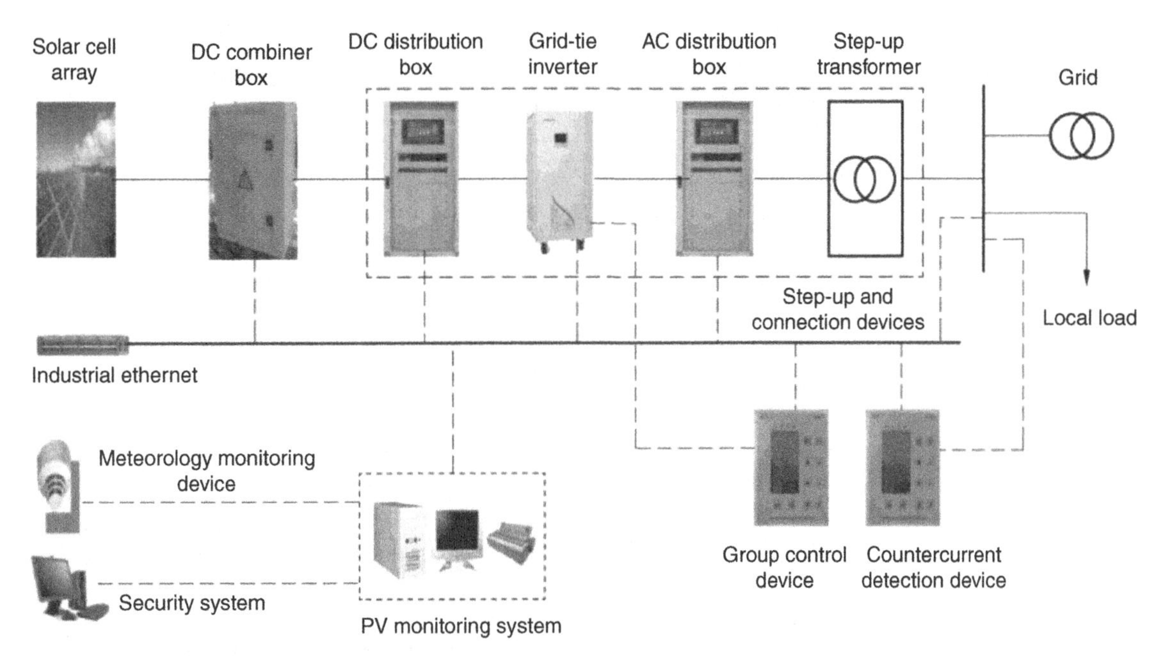

■ **FIGURE 3.2 Structure of grid-connected PV power system.**

PV power systems can be further divided into distributed type and centralized type. The former is a type of DG in a microgrid, in which electricity is directly distributed to users and the surplus or deficit is regulated by the grid. The latter is a PV power system that directly injects electricity to the grid for distribution to users. The structures of these two types are essentially the same.

Figure 3.2 shows the structure of a grid-connected PV power system, mainly comprising solar cell array, DC combiner box, DC distribution box, grid-tie inverter, and AC distribution box. Their functions are the same as those of an independent PV power system. For the off-grid inverter and grid-tie inverter, the similarity is that they both convert direct current to alternating current, and the difference is that the former is the voltage source for U/f output and the latter the current source for P/Q output. In addition, the grid-tie inverter has the following functions: (1) maximum power point tracking (MPPT), that is, it always produces the maximum power when the output voltage and current vary with the cell temperature and solar irradiance; (2) output of current for harmonics suppression to ensure power quality of the grid; (3) automatic tracking of voltage and frequency of the grid in the case of excess power output.

3.3 WIND POWER

Wind energy is a clean renewable energy. Producing electricity is the major way to utilize wind energy, in which the kinetic energy of air in motion is converted to mechanical energy by the rotor and then the mechanical energy

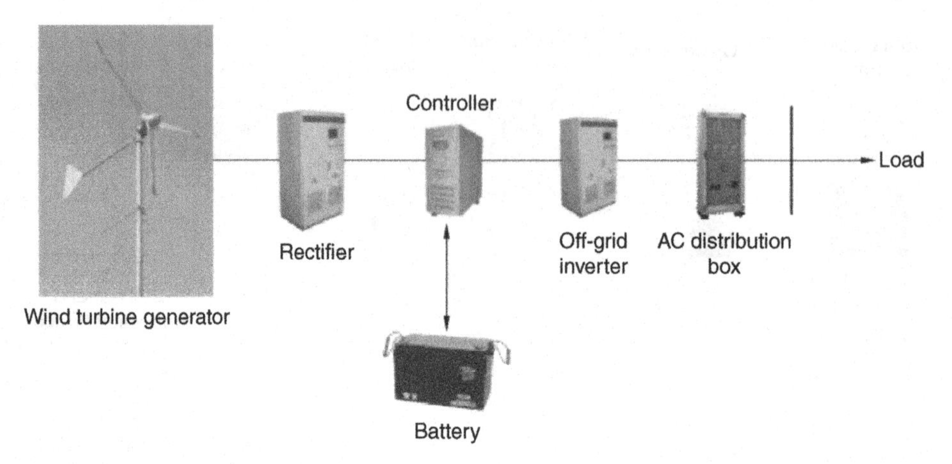

■ FIGURE 3.3 Structure of independent wind power system.

is converted to electricity by the generator. Wind power systems are also divided into independent type and grid-connected type.

3.3.1 Independent wind power system

An independent wind power system is not connected to a traditional electric power system and is mostly deployed in remote off-grid areas to meet local demand. To resolve intermittency of wind-produced electricity, such a system must be provided with ES. Figure 3.3 shows the structure of an independent wind power system, mainly including wind turbine generator, rectifier, controller, battery, off-grid inverter, and AC distribution box. The alternating current from the generator is first converted to direct current by the rectifier to charge the battery and then converted back to alternating current by the inverter to serve AC loads.

3.3.2 Grid-connected wind power system

A grid-connected wind power system is connected to a grid and injects electricity to it. It is the mainstream of wind power systems. Grid-connected wind power systems can be further divided into distributed type and centralized type. The former is a type of DG in a microgrid, in which electricity is directly distributed to users and the surplus or deficit is regulated by the grid. The latter is a wind power system that directly injects electricity to the grid for centralized distribution to users. The structures of the two types are essentially the same.

A grid-connected wind power system may be connected to the grid directly, via inverter or in a hybrid mode, depending on the wind turbine generator.

1. *Direct connection*: Asynchronous generators, including squirrel-cage asynchronous generator and wound induction generator, are directly

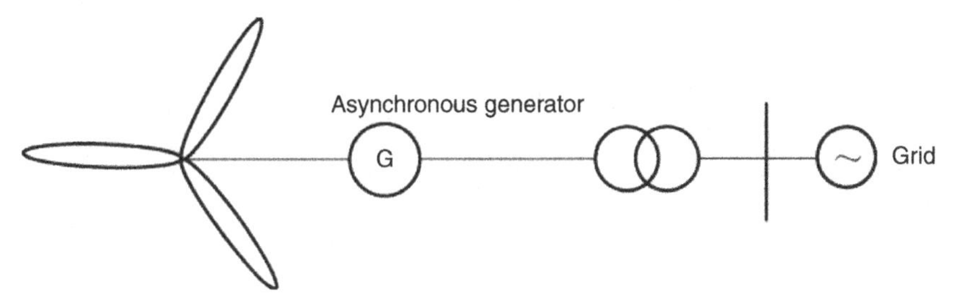

FIGURE 3.4 Schematic diagram of an asynchronous generator directly connected to the grid.

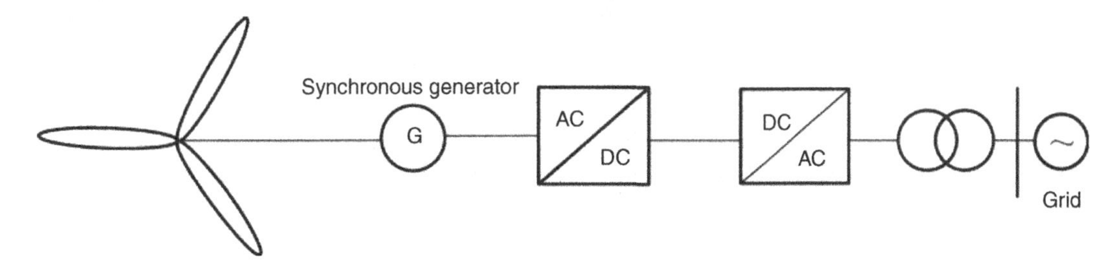

FIGURE 3.5 Schematic diagram of synchronous generator connected to the grid via inverter.

connected to the grid. Figure 3.4 shows the schematic diagram of an asynchronous generator directly connected to the grid. The alternating current produced from the generator is directly injected to the grid.

2. *Connection via inverter*: Synchronous generators, including electrically excited synchronous generator and permanent-magnetic synchronous generator, are connected to the grid via an inverter. Figure 3.5 shows the schematic diagram of a synchronous generator connected to the grid via an inverter. The alternating current produced from the generator is first converted to direct current and then back to alternating current before it is delivered to the grid.
3. *Connection in a hybrid mode*: Doubly fed induction generators are connected to the grid in a hybrid mode, that is, the stator is directly connected to the grid while the rotor is of wound type and connected to the grid via inverter. Figure 3.6 shows the schematic diagram of a doubly fed induction generator connected to the grid in a hybrid mode.

3.4 MICROTURBINE

A micro gas turbine, or simply microturbine, is suitable for various traditional fuels. It makes low noise during operation, and has a much longer life and higher performance than a diesel generator. This small DG system has a short history and first appeared in the United States and Japan. It is advantageous

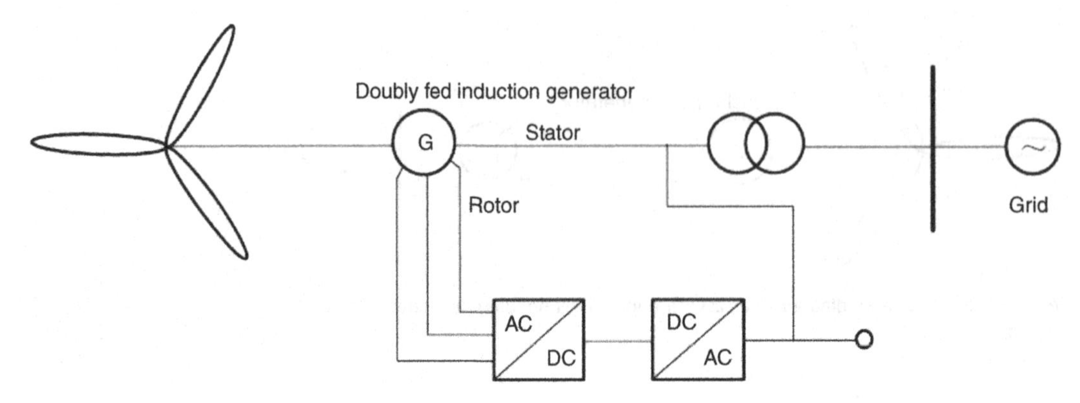

■ **FIGURE 3.6 Schematic diagram of doubly fed induction generator connected to the grid in a hybrid mode.**

because of its small size and weight, reduced maintenance, lower emissions, and high durability, making it popular in the world's energy sector.

The microturbine is an emerging small heat engine, which works in a similar way to the trotting horse lamp popular in China's Tang Dynasty. There is a rotor on the top of the lamp, and when the lamp is lighted, the air inside is heated and moves upward, thus driving the rotor and then the lamp to rotate. Similarly, in a microturbine, the fuel, such as natural gas, methane, petrol, and diesel, is combusted and becomes high-temperature and high-pressure gas, thus driving the rotor to rotate. The unit power ranges from 25 kW to 300 kW. A microturbine consists of a radial flow turbomachinery centripetal turbine and centrifugal compressor, with their rotors arranged back to back on the turbine rotor, an efficient plate regenerator, and a lubrication-free air bearing. The structure is much simpler, as the turbine and generator are integrated. A microturbine may be of a single-shaft or split-shaft structure.

Normally, a microturbine has a speed as high as 50,000–120,000 r/min. A single-shaft microturbine uses a permanent-magnet synchronous generator made of high-energy permanent magnetic materials (such as NdFeB), and the high-frequency alternating current produced is converted to power-frequency alternating current by a power electronic converter. A split-shaft microturbine is connected with the generator through variable-speed gear, and thus power-frequency alternating current can be generated by reducing the speed of the turbine. In a microturbine system, the high-temperature exhaust gas from the generator can be used to preheat the compressed air to be injected to the combustor, thus reducing fuel consumption for combustion and improving energy efficiency. With a lithium bromide absorption refrigerator or heat exchanger, the exhaust gas from the regenerator can be recycled to meet cooling and heat loads.

The components of a microturbine (taking a single-shaft microturbine as an example) are introduced in the subsequent section.

3.4.1 Components of microturbine

The high-speed microturbine is of a simple radial flow design, integrates the circulating reheat technology, and features high reliability, low maintenance cost, low-amplitude vibration, low emissions, and compact structure. It is mainly composed of a single-stage radial compressor, low-emission ring combustor, single-stage radial turbine, compression ratio, and air bearing or bearing with dual lubricating systems.

Figure 3.7 shows the three basic components of a microturbine, namely, the compressor, combustor, and turbine.

Compressor: Air flows through the inlet to enter the compressor, where its pressure and temperature are increased by the blades.
Combustor: The mixture of high-pressure air and natural gas sprayed to the combustor from around is ignited and heated and then expands rapidly.

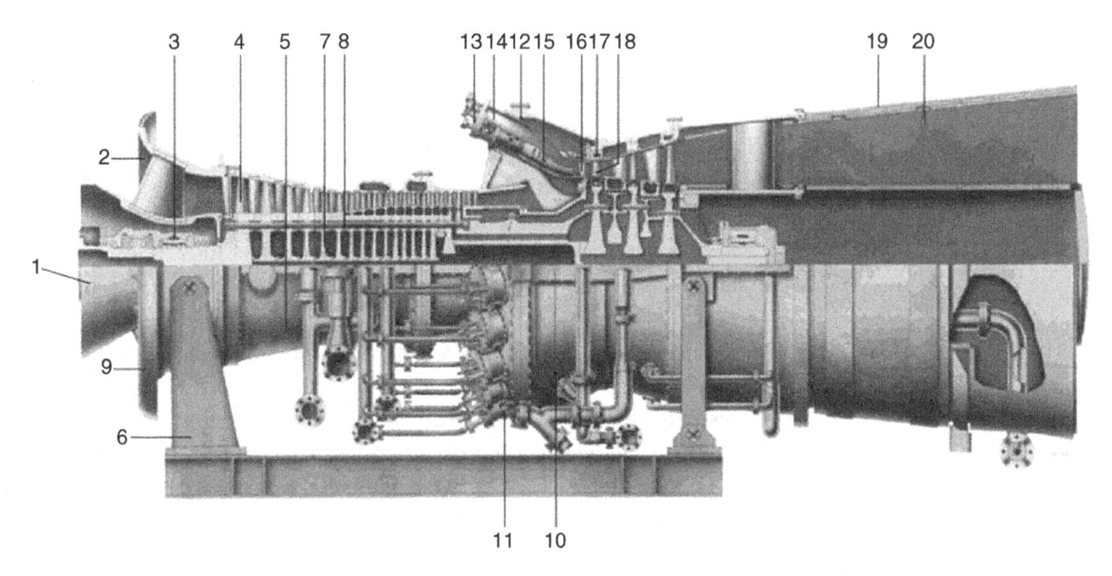

■ **FIGURE 3.7 Profile of a microturbine.** (1) Load coupling. (2) Axial/Radial inlet casing. (3) Radial bearing. (4) Rotor blade of compressor. (5) Mediate casing of compressor. (6) Rigid front support. (7) Disk. (8) Tie-rod. (9) Inlet casing. (10) Horizontal split surface. (11) Front plate of combustor. (12) Counter flow combustor. (13) Fuel distributor. (14) Flame tube of combustor. (15) Impingement cooling structure of combustor transition piece. (16) Stage-one nozzles. (17) Protective ring for stage-one stator blades. (18) Turbine rotor blade. (19) Exhaust diffuser. (20) Thermocouple of outlet casing.

Turbine: The heated, expanded gas enters the turbine and drives blades stage by stage and then the generator rotor to rotate.

3.4.2 High-speed AC generator

The high-speed generator and microturbine are mounted on the same shaft. Thanks to its small size, the generator can be placed inside the microturbine to constitute a compact and high-speed turbine AC generator.

3.4.3 Regenerator

The regenerator increases the efficiency of the microturbine, making the microturbine more competitive than a reciprocating generator. It preheats the air to be injected to the combustor and thus reduces fuel consumption.

3.4.4 Converter system

The high-frequency alternating current produced from the generator must be converted to power-frequency alternating current by the power electronic converter under the control of a microprocessor. The power electronic converter can adjust the speed against load variation of the generator or the grid, or operate as an independent power source. The system facilitates remote management, control, and monitoring, as shown in Figure 3.8.

1. *Soft start and rectifier unit*: During startup of the system, the rectifier bridge operates as an inverter to enable soft start of the permanent-magnet synchronous motor (PMSM); during normal operation, the rectifier bridge functions as a rectifier.
2. *Master inverter power unit*: It mainly consists of an insulated gate bipolar transistor (IGBT) inverter bridge and power filter. It inverts and filters the input DC voltage of 720 V (±40 V) for integration to the grid.
3. *Battery control and management unit*: It controls and manages the smart charge, instantaneous loads, temperature compensation, discharge protection, activation, and self-test of batteries.
4. *Power regulation unit*: It outputs speed signals and adjusts the speed of the microturbine according to the optimal power-speed curve and the relationship between the output power and torque variation of the compressor.

3.4.5 Control system

Figure 3.9 shows the structure of the microturbine control system, in which the regenerator is controlled by adjusting the opening of the active valve

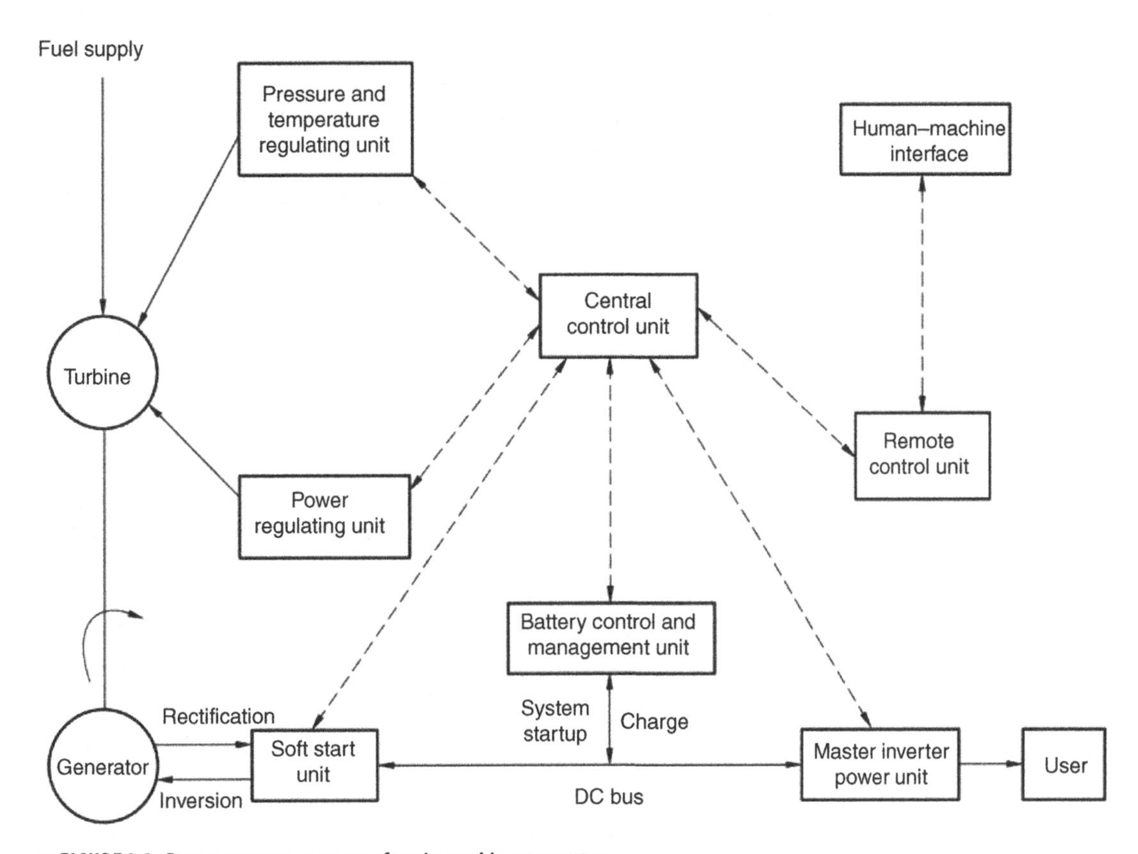

■ **FIGURE 3.8 Power converter system of a microturbine generator.**

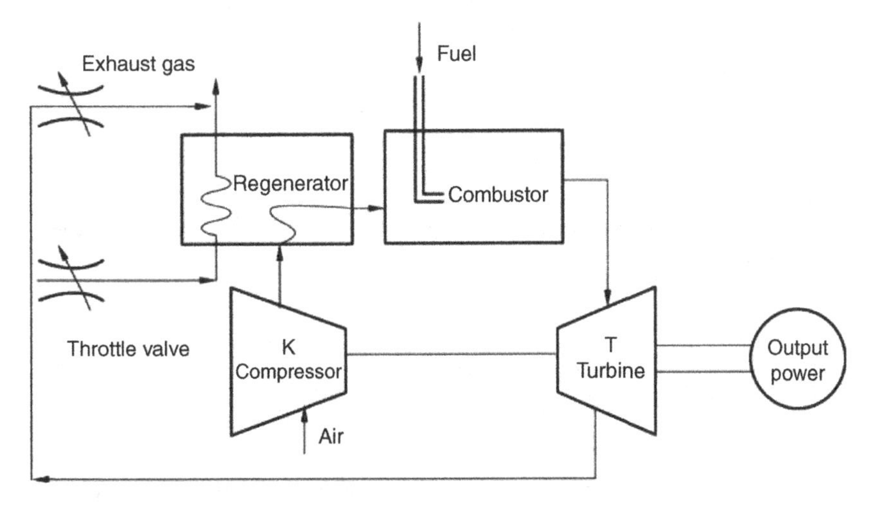

■ **FIGURE 3.9 Structure of microturbine control system.**

and bypass valve to control heat exchange between exhaust gas and clean air (i.e., regenerator effectiveness), so as to control the temperatures of the gas to enter the combustor from the compressor and the exhaust gas and thus improve energy efficiency.

The high-speed single-shaft microturbine is the mainstream type and the most commonly used for small CHP applications. When it is used in a microgrid, its operating status and control methods will affect its dynamic characteristics. Constant active power and reactive power control or *U/f* control may be adopted to ensure stability of the frequency and voltage of the microgrid when operating as an island.

Depending on the system structure, a microturbine may be connected to the grid through an inverter or directly.

1. *Connection to grid through an inverter*: A single-shaft microturbine generator rotates at a very high speed and produces a high-frequency alternating current, and hence, an inverter is required for connection to the grid. Figure 3.10 shows the schematic diagram of a single-shaft microturbine connected to the grid through an inverter. The high-frequency alternating current from the microturbine is first converted to direct current, and then to power-frequency alternating current before it is delivered to the grid.
2. *Direct connection to grid*: For a split-shaft microturbine, the power turbine and gas turbine are mounted on different shafts and connected with the generator through a speed-variable gear. As the generator speed is reduced, it can be directly connected to the grid. Figure 3.11

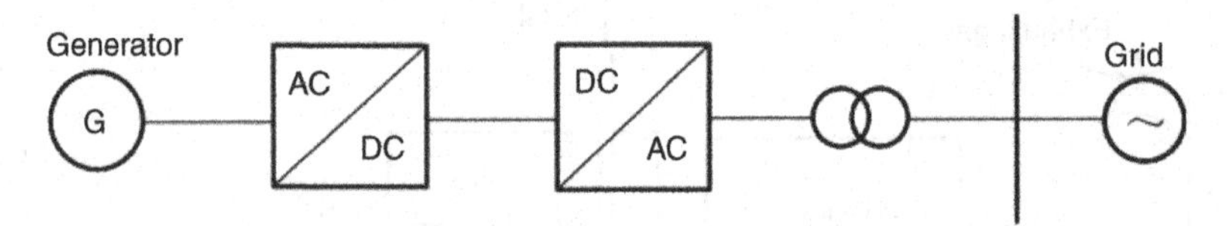

■ **FIGURE 3.10 Schematic diagram of single-shaft microturbine connected to the grid through inverter.**

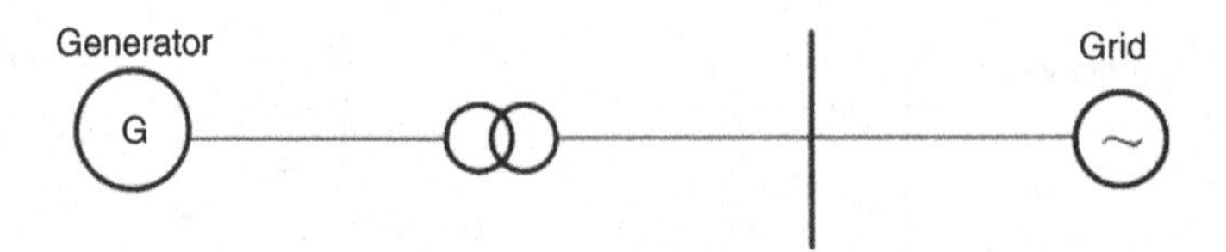

■ **FIGURE 3.11 Schematic diagram of split-shaft microturbine directly connected to the grid.**

shows the schematic diagram of a split-shaft microturbine directly connected to the grid.

3.5 OTHER TYPES OF DGS

In addition to PV, wind, and microturbine discussed earlier, there are some other types of DGs that currently are not widely used due to immature technologies, including geothermal, biomass, and ocean energy. The following gives an outline of these energy forms.

3.5.1 Geothermal energy

Producing electricity is the major means to utilize geothermal energy. Similar to thermal power generation, in geothermal generation, the thermal energy of steam is converted to mechanical energy in the turbine to drive the generator to produce power. The difference is that geothermal generation does not require a huge boiler and uses geothermal sources instead of fuels. Geothermal power systems may be of flash steam type or binary cycle type, depending on the temperature of the geothermal source. Moreover, total flow generation and dry hot rock generation are currently under study.

3.5.2 Biomass energy

In the broad sense, biomass is the organic matter generated from the photosynthesis process of plants and therefore, its energy is converted from solar energy. Biomass energy may come from agricultural, forestry, and fisheries resources such as crops, wood, and seaweed, or industrial organic waste such as pulp waste, black liquid from paper making, and residue from alcoholic fermentation, or common municipal waste such as household garbage and paper scraps, or residual sludge from sewage treatment plants. Biomass power generation comes in three forms: (1) biomass combustion, (2) biomass gasification, and (3) methane combustion.

1. *Biomass combustion*: In biomass combustion electrification, the biomass is combusted in the boiler, the hot gas heats the heating surface of the boiler to produce high-temperature high-pressure steam, which expands in the gas turbine and drives the turbine to produce electricity. Generally, the biomass needs to be processed before combustion for a higher efficiency.
2. *Biomass gasification*: In biomass gasification electrification, the biomass is first converted to gas fuel by thermal chemical reaction, and then the gas fuel is purified and directly injected to the boiler, internal combustion engine, and combustor of the gas turbine to produce electricity.

3. *Methane combustion*: Methane combustion electrification emerges following the improved use of methane. The methane is used for the generator and other devices to produce power and heat.

3.5.3 Ocean energy

Ocean energy, a kind of renewable energy generated from the motion of seawater, mainly includes thermal gradient energy, tide energy, wave energy, tidal current energy, sea current energy, and salinity gradient energy. Accordingly, electrification technologies using ocean energy include tide electrification, sea current electrification, tidal current electrification, wave electrification, and thermal gradient electrification.

1. *Tide electrification*: Tide is a result of the relative motion of the Earth, Moon, and Sun, and thus is somewhat predictable. Tidal power is generated using the potential energy produced from the rise and fall of tides. Among all ocean energy technologies, it is the most mature and widely used type. The principle is as follows: a dam is built in an appropriate location. During the rise of tides, seawater flows to the dam reservoir and drives the hydraulic turbine to rotate to produce electricity; and during the fall of tides, the seawater flows back to the sea and also drives the turbine to produce electricity. As a result, the water turbine shall be able to rotate in the same direction irrespective of the direction of the water flow.
2. *Sea current electrification*: Sea current is an electrification form using the kinetic energy of sea currents and tidal currents moving in a given direction. The generation facility, similar to a wind turbine, is called "underwater windmill." Facilities to convert the kinetic energy of sea currents to electricity may be of a propeller type, symmetrical airfoil and perpendicular shaft wheel type, and parachute type or magnetic flow type. The magnetic flow type works like this: when the sea currents containing large amounts of electric ions flow through the magnetic field, induced electromotive force will be generated and electricity is produced.
3. *Tidal current electrification*: Tidal currents are periodic seawater currents resulting from tides, and are similar to tides. They are weaker in the deep sea and stronger when getting closer to the shore, especially at the narrow entrance to a gulf or a narrow strait or waterway formed after separating from the land, where the water flows very fast. As such, the geographical distribution of fast-moving tidal currents is coincidental with that of large tidal ranges. To obtain energy from tidal currents, water turbine that can rotate in the same direction when the tidal currents flow in opposite directions is used. Different from tide electrification, tidal current electrification requires no dam or levee,

and the tidal currents directly pass through the turbine by gravity to produce electricity.

4. *Wave electrification*: Wave electrification may be of floating type or stationary type. The principle is that waves move the water in the air chamber upward and downward and then the air above the installation moves back and forth, thereby driving the turbine to rotate and produce electricity.
5. *Thermal gradient electrification*: This is an electrification form using the temperature difference between surface water (at 26–27° C as most of solar radiation energy is converted to thermal energy) and deep water (at 1–6° C). It comes in two types, namely, closed circulating type and open circulating type. For the former type, the medium (ammonia or Freon) is evaporated by the heat of surface water (13–25° C) in the evaporator, enters the turbine, and drives the generator to rotate. Then the medium is cooled to liquid state by deep water in the condenser and is again sent to the evaporator by the circulating pump, in the same circulating sequence as in thermal power generation. For the latter type, the low-pressure surface water is sent to the evaporator with a vacuum pump, where a small portion of water is evaporated and the rest is sent back to the sea, and under the action of the vacuum pump, the evaporated water drives the turbine to produce electricity. After that, the water enters the condenser and is cooled and desalinated for drinking and industrial use.

3.6 ENERGY STORAGE

ES technologies solve the imbalance between supply and demand. In distributed generation, they solve the intermittency and uncertainty of output and load shifting. They also enable a number of other functions such as black-start, power quality adjustment and control, and stability.

ES technologies include physical, electromagnetic, electrochemical, and phase-change forms. Physical ES technologies include pumped storage, compressed air ES (CAES) and flywheel ES; electromagnetic ES includes superconducting magnetic ES (SMES), supercapacitor, and high-energy-density capacitor; electrochemical ES includes various types of batteries such as lead-acid battery, nickel-hydrogen battery, nickel-cadmium battery, lithium-ion battery, sodium sulfur battery, and flow battery; phase-change form includes ice thermal storage.

3.6.1 Pumped storage

Pumped storage is a physical ES form. To use this form, an upstream reservoir and a downstream reservoir are required at the storage station. During off-peak hours, the pumped storage device operates as a motor to pump the water from the downstream reservoir to the upstream; and during peak hours, the device operates to produce electricity using the water in the upstream reservoir. A pumped storage station could have a capacity of energy for release for a few hours to a few days, with an efficiency ranging from 70% to 85%. Among all ES forms, this is the most widely used form in power systems, mainly for load shifting, frequency and phase modulation, emergency reserve, black-start, and system reserve. It can also improve the efficiency of thermal and nuclear power stations.

3.6.2 Compressed air ES

CAES is a physical ES form realized by means of gas turbine power generation and is intended for load shifting. It works like this: during off-peak hours, the excess power of the grid is used to compress the air, which is then stored in a typical 7.5 MPa sealed container and released during peak hours to drive the turbine to generate power. During the power generation, 2/3 of the gas is used to compress air, reducing gas consumption by 1/3, and the gas consumption is 40% less than that of a conventional gas turbine. CAES makes cold start and black-start possible, and can respond immediately. It is mainly for power regulation, load balance, frequency modulation, and distributed ES and reserve of power systems.

3.6.3 Flywheel ES

Flywheel is a physical ES form. A flywheel ES system is composed of a high-speed flywheel, bearing support system, motor/generator, power converter, electronic control system, and optional equipment including vacuum pump and backup bearing. During off-peak hours, the grid energizes the flywheel to rotate fast, thus realizing conversion from electricity to mechanical energy; during peak hours, the high-speed flywheel works as a prime mover for the generator to produce electricity and output current and voltage via the power converter, thus realizing conversion from mechanical energy to electricity. The flywheel is mainly used for uninterrupted power supply (UPS), emergency power system (EPS), load shifting, and frequency control.

3.6.4 Superconducting magnetic ES

SMES, an electromagnetic ES form, uses superconducting coils to store magnetic energy and requires no energy conversion for power transmission. It has such advantages as quick response (within several ms), high conversion efficiency, high specific capacity/power, and real-time large-capacity energy exchange with and power compensation for the electric power system. It is mainly used for voltage support, power compensation, and frequency regulation of TD networks to improve system stability and transmission capacity.

3.6.5 Supercapacitor

A supercapacitor, an electromagnetic ES form, is developed based on the electromagnetic double-layer theory and can provide very high pulse power. In the charging process, the charges on the surface of an ideally polarized electrode will attract the ions of opposite polarity in the electrolyte to the electrode surface, forming an electric double-layer and thus an electric double-layer capacitor. As the spacing between the two layers is very small, generally less than 0.5 mm, and the electrode is of a special structure, the surface area of the electrode increases by tens of thousands of times, creating an extremely large capacitance. However, due to the low voltage withstand level of the dielectric medium and occurrence of current leakage, the amount and duration of energy storage are limited, and supercapacitors must be series-connected to increase the volume of the charge and discharge control circuit and the system. The supercapacitor is mainly used for short-time and high-power load shifting, for reliable supply following a voltage drop and instantaneous disturbance, and in situations of high peak power, for example, support for startup and dynamic voltage recovery of a high-power DC motor.

3.6.6 Battery

Battery is a chemical ES form and comes in many types, mainly lead-acid battery, nickel-cadmium battery, lithium-ion battery, sodium sulfur battery, and vanadium redox flow battery.

1. *Lead-acid battery*: Its lifetime will be reduced when working at a high temperature. Similar to a nickel-cadmium battery, it has a low specific energy and specific power, but is advantageous because of its low price and cost, high reliability, and mature technology and has been widely used in electric power systems. However, it has a short lifetime and causes environmental pollution during manufacture. It is mainly used as the power source for closing of circuit breakers during system operation, and an independent power source for relay protection, driver motor, communication, and emergency lighting in the event of failure of power plants or substations.
2. *Nickel-cadmium battery*: It has a high efficiency and long lifetime, but the capacity decreases as time goes by, and the charge retention needs

to be enhanced. Furthermore, it has been restricted by the EU due to heavy metal pollution. It is rarely used in electric power systems.

3. *Lithium-ion battery*: It has a high specific energy and specific power, little self-discharge, and causes no pollution. However, due to the influence of the process and difference in ambient temperature, the system indices are more often worse than those of a cell, and in particular, the lifetime is several times or even more than 10 times shorter than that of a cell. What is more, integration of a high capacity is very difficult and the cost for manufacture and maintenance very high. In spite of this, the lithium-ion battery is expected to be widely used in DG and the microgrid thanks to advancements of technologies and reduction of costs.
4. *Sodium sulfur battery*: Owing to its high energy density, its size is just 1/5 of a lead-acid battery while the efficiency is up to 80%, contributing to convenient modular design, transportation, and installation. It can be installed by stage according to the intended purpose and capacity, and suits urban substations and special loads. It is a promising ES technology for DG and the microgrid in improving the system stability, shifting loads, and maintaining power supply in an emergency.
5. *Flow battery*: Flow battery features slight electrochemical polarization, 100% discharge, long lifetime, and rated power independent of rated capacity. The capacity can be increased by adding electrolyte or increasing the concentration of the electrolyte. The storage form and pattern can be designed according to the location. It is a promising ES technology for DG and the microgrid in improving the system stability, shifting loads, and maintaining power supply in an emergency.

Chapter 4

Control and operation of the microgrid

There are two types of microgrids: (1) independent microgrid and (2) grid-connected microgrid. The control of an independent microgrid is complicated, involving steady-state control, dynamic control, and transient control, while a grid-connected microgrid only involves steady-state control.

4.1 THREE-STATE CONTROL OF INDEPENDENT MICROGRID

Independent microgrids are mainly deployed in remote areas that are not covered by a utility grid, such as an island, a mountainous area, or a village, and the distribution system of the grid is powered by diesel generators (or gas turbines). The distributed generation (DG) capacity is close to or higher than the capacity of the distribution system, making the system a high-penetration independent microgrid.

An independent microgrid is hard to control due to the small capacity of the distribution system and the high penetration of DGs. For such a microgrid, steady-state constant-frequency and constant-voltage control, dynamic generator tripping and load shedding control, and transient fault protection can ensure its stability. Figure 4.1 is a schematic diagram of the three-state control system of an independent microgrid. In the system, an intelligent data collection terminal is arranged on each node, which collects the current and voltage of the node and sends the data to the microgrid control center (MGCC) via the Internet. The MGCC is made of the three-state stability control system, including the centralized protection and control equipment, dynamic stability control equipment, and steady-state energy management system. According to the dynamic characteristics of voltage and frequency, the control system identifies stable voltage and frequency zones, as detailed in Figure 4.2:

> *Zone A*: the deviation from the rated voltage and frequency is within the normal fluctuation range and does not affect the power quality.

Microgrid Technology and Engineering Application

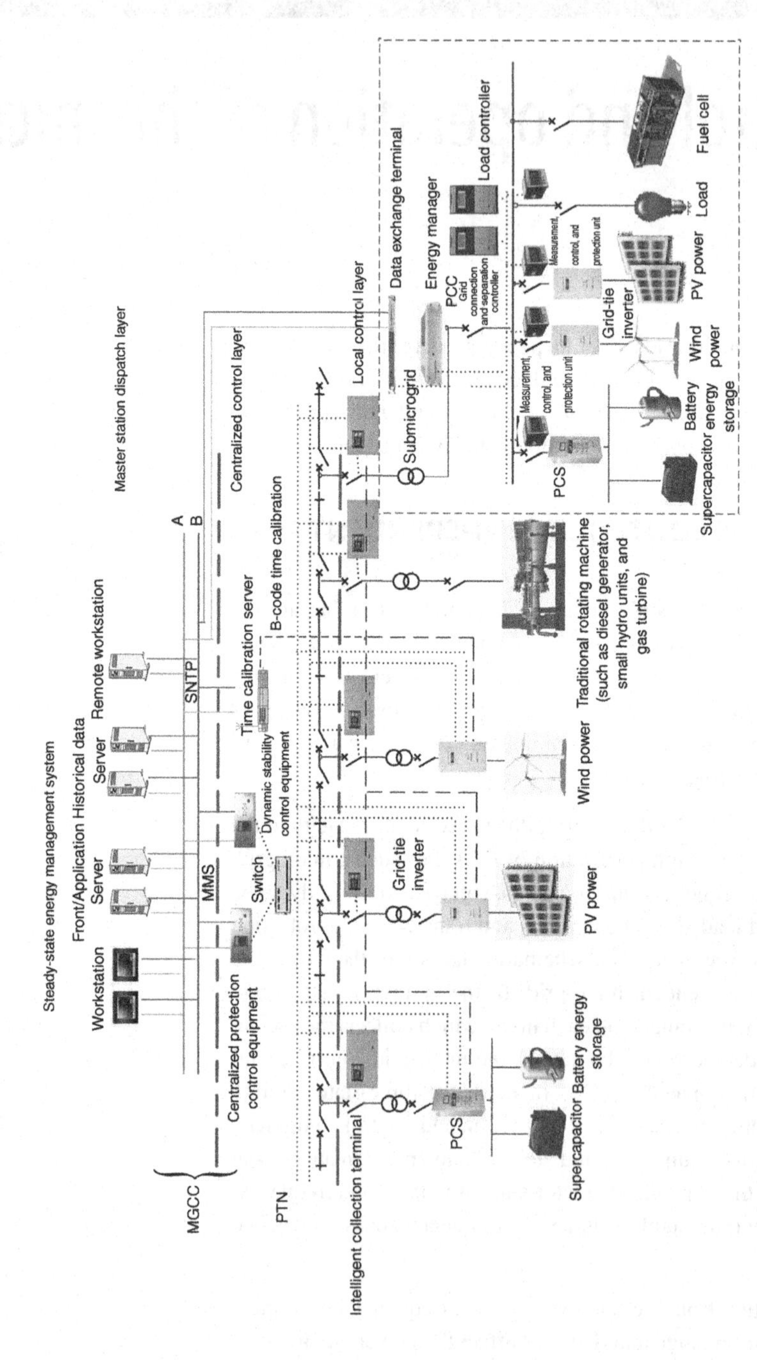

■FIGURE 4.1 Schematic diagram of three-state control system of an independent microgrid.

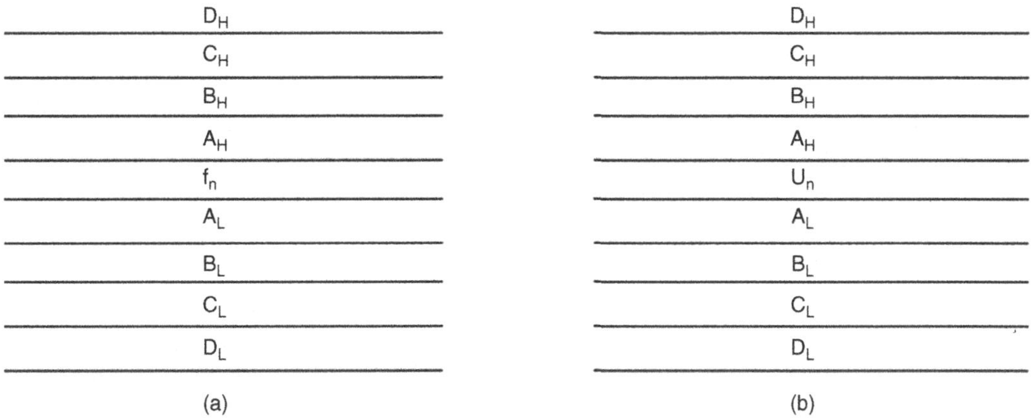

(a)	(b)
D_H	D_H
C_H	C_H
B_H	B_H
A_H	A_H
f_n	U_n
A_L	A_L
B_L	B_L
C_L	C_L
D_L	D_L

FIGURE 4.2 Classification of stable voltage and frequency zones. (a) Stable voltage zones and (b) stable frequency zones.

Zone B: the deviation is slightly beyond the allowable range, and can quickly restore to the range of Zone A through energy storage (ES) regulation.
Zone C: the deviation is far beyond the allowable range, and generator tripping and load shedding are required for maintaining system stability.
Zone D: the deviation is beyond the controllable range, and the grid is subject to a large disturbance (a fault, for example), necessitating quick clearing of the fault to resume system stability.

4.1.1 Steady-state constant-frequency and constant-voltage control

In steady operation, an independent microgrid is not subject to large disturbance or significant load variation, the outputs of diesel generators and DG are balanced with load, the voltage, current and frequency stay around the average values, and the voltage and frequency are in the range of Zone A. The steady-state energy management system executes steady-state constant-frequency and constant-voltage control to balance DG outputs, and monitors and analyzes in real-time the voltage U, frequency f, and power P of the system. If loads vary little, and U, f, and P are within the normal range, the system checks DC outputs and controls charge and discharge of ES to balance DC outputs. See Figure 4.3 for the control flowchart.

1. In the case of excess DG outputs, the energy management system checks the state of charge (SOC) of the ES system. If the SOC has reached the upper limit, the ES device has been fully charged, and the

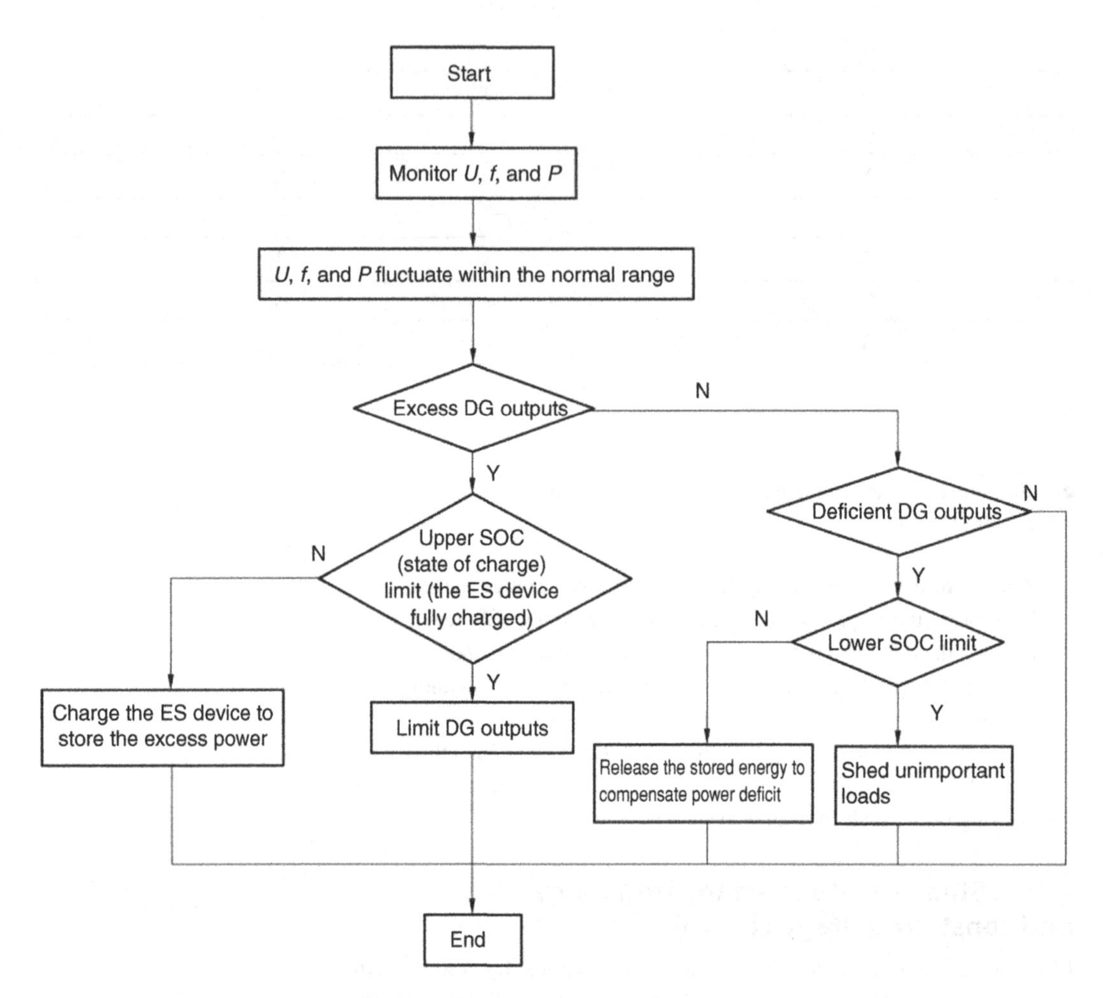

■ **FIGURE 4.3 Steady-state constant-frequency and constant-voltage control.**

DG outputs will be limited; while if the SOC is below the upper limit, the ES device is charged to store excess power;

2. In the case of deficient DG outputs, the energy management system checks the SOC of the ES device. If the SOC has reached the lower limit, unimportant loads are shed and discharge is stopped; if the SOC is above the lower limit, the stored energy will be discharged;
3. In the case of balance between DG outputs and loads, no regulation of the ES, DG, and loads is necessary.

As stated earlier, maintaining balance between intermittent DG outputs and load demand through control of ES charge and discharge, DG outputs and loads could keep an independent microgrid in steady operation.

4.1.2 Dynamic generator tripping and load shedding control

System frequency is one of the most important indicators for power quality. To ensure normal operation, it must maintain around 50 Hz. Otherwise, both the generators and loads will be adversely affected, and in the worst case, frequency collapse may occur. Load fluctuation will lead to grid frequency fluctuation. There are three types of loads in terms of the fluctuation pattern: (1) random loads featured with a low fluctuation amplitude and short interval (generally within 10 s); (2) loads featured with a high fluctuation amplitude and long interval (generally 10 s to 30 min), such as a furnace and motor: (3) and loads fluctuating slowly but constantly as a result of patterns of people's daily activities. When suffering a dynamic disturbance caused by load fluctuation, the system should be capable of reestablishing and maintaining stability.

In a traditional large grid system, the frequency deviation resulting from load fluctuation is limited by regulating the system's frequency. Frequency deviations caused by low-amplitude, short-interval load fluctuation are generally regulated by the speed governor of the generator, known as the primary regulation. Frequency deviations caused by high-amplitude, long-interval load fluctuation cannot be limited within the allowable range solely by the speed governor, and require frequency modulation by a frequency modulator, known as the secondary regulation.

An independent microgrid does not have such speed governor for primary regulation or frequency modulator for secondary modulation. And when subject to a dynamic disturbance due to load fluctuation, it is incapable of reestablishing stability. As such, dynamic generator tripping and load shedding control by the dynamic stability control system is required for dynamic stability of the microgrid.

As shown in Figure 4.1, the intelligent terminals on individual nodes collect the measured data of the node and send the data to the dynamic stability control system for real-time monitoring and analysis of the voltage U, frequency f, and power P. If the loads fluctuate significantly and U, f, and P are beyond their normal range, the dynamic stability control system checks DG outputs and coordinately controls the ES, DG, loads, and reactive compensation equipment. See Figure 4.4 for the control flowchart.

1. Sudden increase of loads will cause power deficit and drop of voltage and frequency. If f drops to Zone B_L, the stored energy is released to provide the power deficit if the disturbance lasts for less than 30 min; if longer than 30 min, unimportant loads are shed as well to protect the ES; if f drops to Zone C_L, unimportant loads are directly shed due

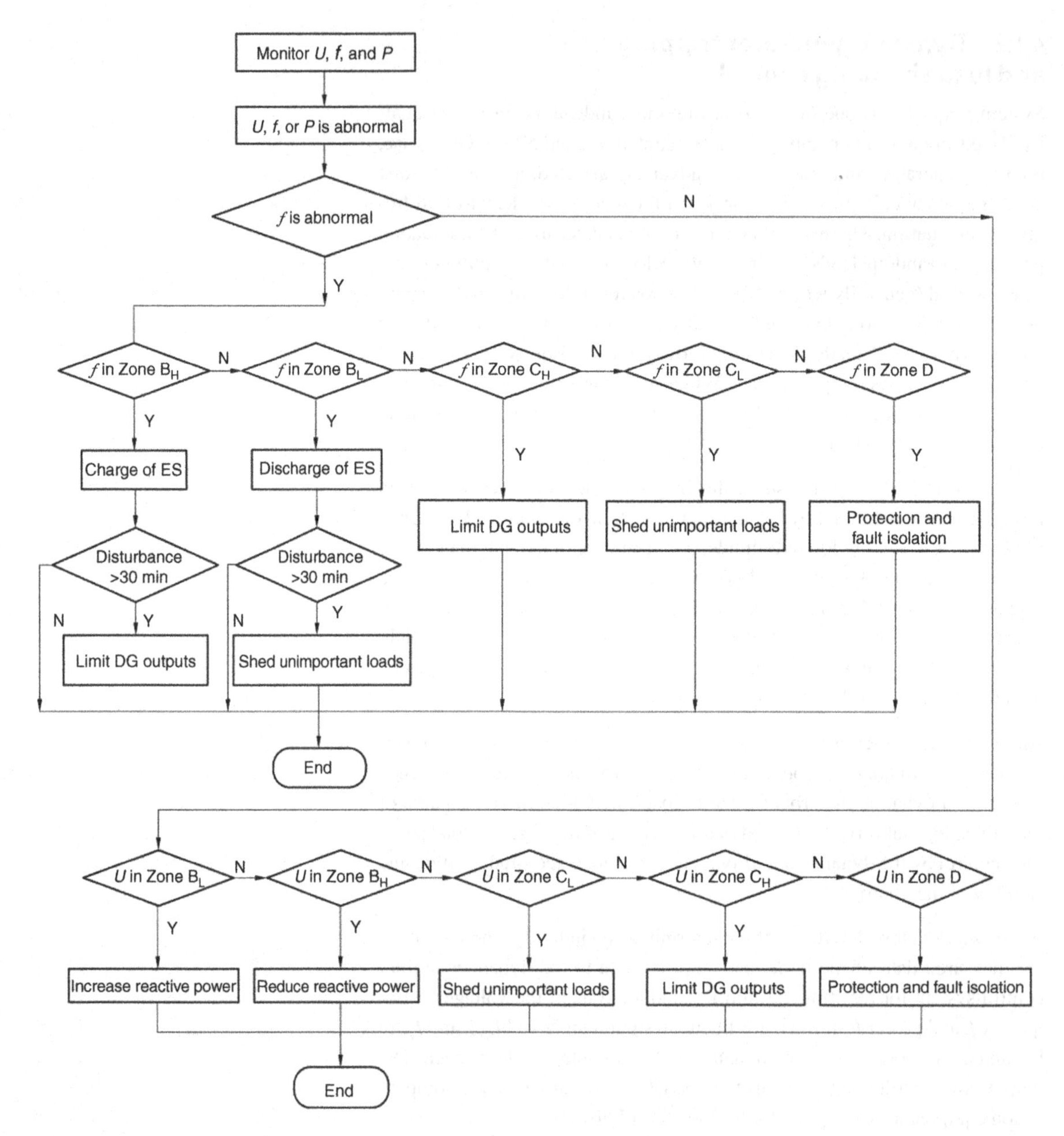

FIGURE 4.4 Dynamic under-frequency load shedding control.

to the large fluctuation. If U drops to Zone B_L, reactive compensation equipment is used to provide the power deficit; if U drops to Zone C_L, unimportant loads are shed.

2. Sudden drop of loads will cause power surplus and rise of voltage and frequency. If f rises to Zone B_H, the surplus power is stored in the ES

if the disturbance lasts for less than 30 min; if longer than 30 min, DG outputs are limited as well. If *f* rises to Zone C_H, DG outputs are directly limited. If *U* rises to Zone B_H, the reactive power is reduced to regulate the voltage; if *U* rises to Zone C_L, unimportant loads are shed. If the disturbance lasts for more than 30 min, ES does not participate in regulation, so as to keep its SOC within the range from 30% to 70% for dynamic regulation in the case of load fluctuation with a small amplitude and short interval.

3. Faults: when the voltage and frequency become abnormal following a fault, and tripping generators and shedding loads cannot restore the system's stability, protection and fault isolation measures will be taken. This is transient fault protection as detailed in Section 4.1.3.

As stated, through the control of ES charge and discharge, DG outputs, and loads in response to load fluctuation to maintain dynamic balance of voltage and frequency, the microgrid can keep steady operation.

4.1.3 Transient fault protection

Transient stability of an independent microgrid refers to the microgrid's ability to resume a new or the original steady state following a large disturbance such as a short-circuit fault and a sudden breakage. If the fault is not cleared immediately, the microgrid will be unable to serve the loads and will lose frequency stability, thus causing frequency collapse and blackout of the entire system.

To ensure transient stability of an independent microgrid, the faults on the distribution system of the main grid, such as faults on the line, bus, step-up transformer, and step-down transformer must be cleared quickly by relay protections.

Based on the characteristics of an independent microgrid when a fault occurs, fast decentralized collection and centralized processing are combined, and the centralized protection and control device is provided instead of conventional distribution network protection to realize quick self-healing. The device has the following major functions:

1. Automatically open or close the grid-tie switch based on the voltage and current of each node of the distribution system after a fault occurs, allowing for fault isolation, network restructuring, uninterrupted power supply, and thus high supply reliability.
2. Quickly identify optimal energy path and reduce or eliminate overloading, thereby realizing basic balance between generation and consumption.

3. Work on a differential protection basis, that is, the integration of DG on any node within the protected area will not affect the protection effects and ratings.
4. Directly locate faults, eliminating the time delay of busbar automatic transfer switch caused by upper-level and low-level coordination, allowing for quick resumption of power supply and improving power supply quality.

Transient fault protection greatly speeds up fault identification, shortens the duration of blackout and improves system stability. For details, see Chapter 5 "Protection of Microgrid."

Transient fault protection depending on quick fault clearing techniques, together with steady-state constant-frequency and constant-voltage control and dynamic generator tripping and load shedding control, realizes energy balance and control of an independent microgrid at three states and ensures security and stability of the microgrid.

4.2 INVERTER CONTROL

4.2.1 Grid-tie inverter control

A grid-tie inverter is used for unidirectional energy flow from DG to the grid. In grid-connected operation, the inverter obtains reference voltage and frequency from the grid. In islanded operation, the inverter serves as the slave power source, obtains reference voltage and frequency from the main power source, is under *P*/*Q* control and controls its active and reactive outputs according to orders from the MGCC.

4.2.2 Power converter system control

The power converter system (PCS), a bidirectional inverter interconnecting the ES and the grid, allows for bidirectional energy flow between the two. It executes *P*/*Q* control over the ES and regulates the power outputs of DGs. In islanded operation, it works as the main power source to execute *U*/*f* control, provide voltage reference for the microgrid, and enable black start of the microgrid. See Figure 4.5 for the block diagram of the PCS.

1. *P*/*Q control*: The PCS can control its active input/output and reactive input/output according to the orders from the MGCC, allowing for bidirectional regulation of active power and reactive power.
2. *U*/*f control*: The PCS can realize constant-voltage and constant-frequency output control according to the orders from the MGCC, and serve as the main power source to provide voltage and frequency reference for DGs.

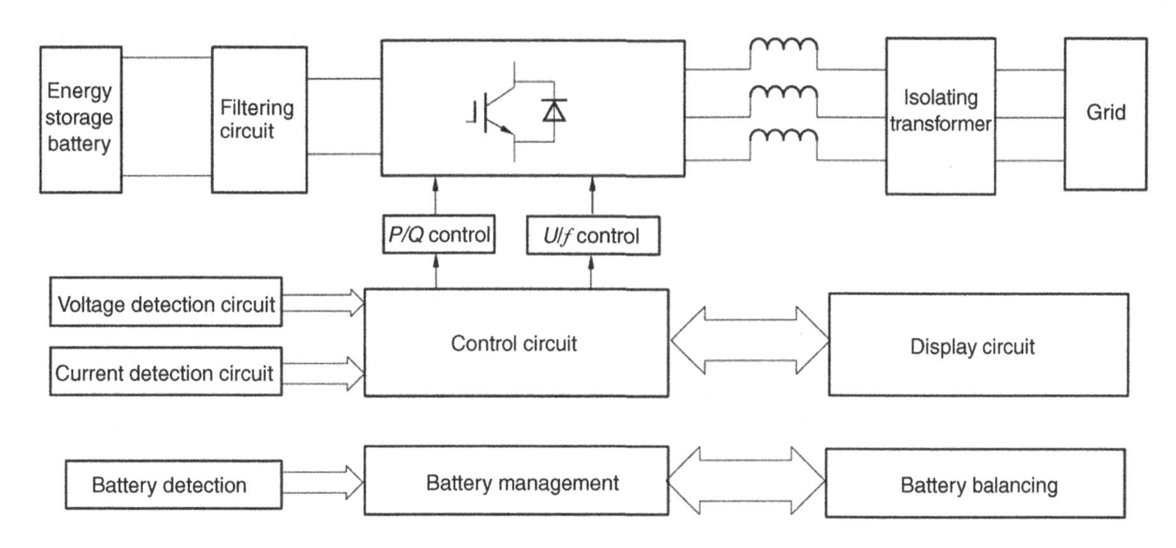

■ **FIGURE 4.5 Block diagram of PCS.**

3. *Battery management system (BMS)*: The BMS mainly serves to monitor the status and estimate the SOC of batteries, to prevent over-charge and over-discharge, ensure safe use, extend the service life, and improve the efficiency of batteries. Its main functions are as follows:
   - **a.** Check the SOC, or residual energy of batteries, to protect the battery from damage due to over-charge or over-discharge.
   - **b.** Dynamically monitor the battery status, and collect the terminal voltage, charge/discharge current, and temperature of each cell in the battery and the total voltage of the battery during the charge/discharge process, to prevent over-charge or over-discharge; identify failed cells to ensure reliability and efficiency of the battery, and makes it possible to build a model for estimating residual energy of batteries.
   - **c.** Making balance between cells to ensure that the cells in a battery are evenly charged.

4.3 GRID CONNECTION AND SEPARATION CONTROL

Apart from the grid-connected operation and islanded operation discussed in Chapter 2, a microgrid may also be in a transition mode, including transition from grid-connected to islanded, from islanded to grid-connected, and transition to shutoff.

In grid-connected operation, the grid provides the power deficit or absorbs the power surplus of DG to maintain energy balance within the microgrid. In this mode, optimization and coordination are required for maximizing the energy efficiency, that is, to utilize DG outputs as much as possible for best economy of the microgrid provided that operation constraints are met.

1. Transition from grid-connected mode to islanded mode
   When a fault occurs on the distribution network or the microgrid is scheduled to operate in islanded mode, the microgrid switches to this transition mode. First, the circuit breaker at the point of common coupling (PCC) is opened, and the islanding protection of DG inverters act and DG quits service, leading to temporary power interruption. Then unimportant loads are shed and the main power source switches to *U*/*f* control from *P*/*Q* control to ensure continuous supply to important loads. After DGs resume operation, depending on their output, some loads shed previously are reconnected. Thus the microgrid enters the islanded mode. In islanded mode, energy balance and voltage and frequency stability within the microgrid are achieved by the control system, and on this condition, the power quality is improved and DG output is utilized as practical as possible.
2. Transition from islanded mode to grid-connected mode
   During islanded operation, when the microgrid detects that the distribution network resumes operation or receives an order from the energy management system to end the intentional islanding, the microgrid prepares for reconnection to grid and resumption of supply to loads shed previously. At this time, if the microgrid meets the voltage and frequency conditions for grid connection, it will switch to the transition mode. First, the circuit breaker at the PCC is closed to resume power supply to loads. Then the main power source of the microgrid is switched to *P*/*Q* control from *U*/*f* control. Thus, the microgrid enters the grid-connected mode.
3. Transition to shutoff
   When a fault occurs on the DG or other equipment within the microgrid, causing failure of control and coordination of DG output, the microgrid will be shut off for maintenance.

A microgrid constantly transits between these operation modes, most often between grid-connected mode and islanded mode.

4.3.1 Grid connection control

4.3.1.1 *Conditions for grid connection*

Figure 4.6 shows the system diagram and phasor diagram for a microgrid connected to the distribution network.

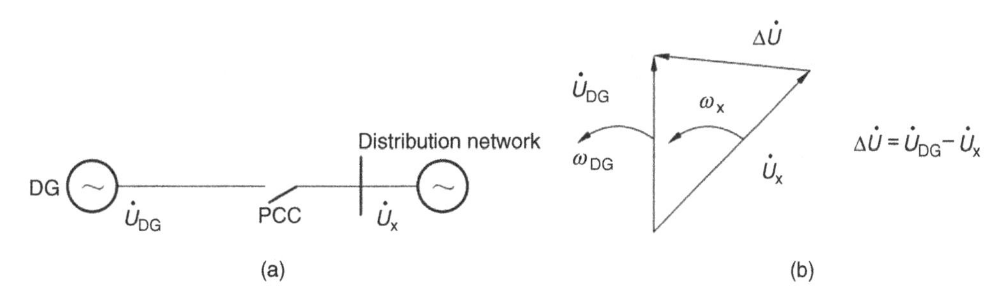

FIGURE 4.6 System diagram and phasor diagram for a microgrid connected to the distribution network. (a) System diagram and (b) phasor diagram.

U_x represents the voltage on the distribution network side and U_{DG} the operation voltage of the microgrid in islanded mode. The ideal condition for connecting a microgrid to a distribution network is:

$$f_{DG} = f_x \text{ or } \omega_{DG} = \omega_x (\omega = 2\pi f) \tag{4.1}$$

$$\dot{U}_{DG} = \dot{U}_x \tag{4.2}$$

The phase angle difference between $\dot{U}_{DG}$ and $\dot{U}_x$ is zero, that is, $|\delta| = |\arg(\dot{U}_{DG} / \dot{U}_x)| = 0$.

When Eqs (4.1) and (4.2) are met, the rush current arising from the closing of the circuit breaker is zero, and the DG is synchronized with the distribution network. In fact, it is hard and unnecessary to meet the ideal condition in Eqs (4.1) and (4.2), as long as the rush current is small enough to cause no problem. The actual synchronization criteria are as follows:

$$|f_{DG} - f_x| \le f_{set} \tag{4.3}$$

$$|\dot{U}_{DG} - \dot{U}_x| \le U_{set} \tag{4.4}$$

where f_{set} represents the setting value of frequency difference between the distribution network and microgrid, and U_{set} the setting value of voltage difference between the two.

4.3.1.2 Grid-connection logic

There are two grid-connection logics: no voltage and synchronization:

1. *No-voltage grid-connection*: The logic applies to the scenario when a microgrid is shut off, the ES and DG are not in operation, the distribution

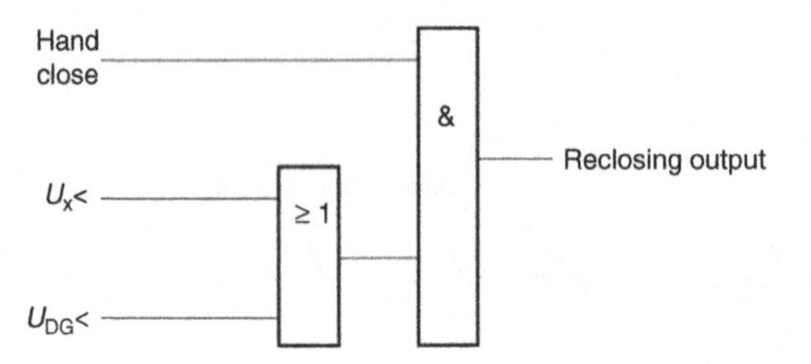

■ **FIGURE 4.7 No-voltage grid-connection logic.**

network serves the loads, and the circuit breaker at the PCC is ready for no-voltage grid connection. See Figure 4.7 for the schematic diagram of a no-voltage grid-connection logic, where the circuit breaker is closed manually or through remote control, and "U_x <" and "U_{DG} <" represent zero voltage.

2. *Synchronization-based grid-connection*: This logic applies to the scenario when the grid resumes to normal operation or an order to end intentional islanding is received from the energy management system of the microgrid, the microgrid and grid will be checked for synchronization. If they are synchronized, the circuit breaker at the PCC is closed, an order to shift to grid-connected mode is sent, and the ES ceases to output power and switches to *P*/*Q* control from *U*/*f* control. After the circuit breaker is closed, the microgrid is reconnected to the grid.

See Figure 4.8 for the schematic diagram of synchronization-based grid connection logic, where "U_x >" and "U_{DG} >" represent nonzero voltage, and a time delay of 4 s is for confirmation of voltage stability.

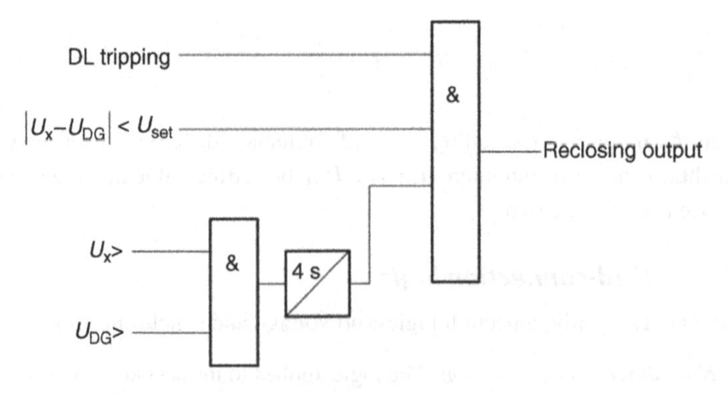

■ **FIGURE 4.8 Synchronization-based grid-connection logic.**

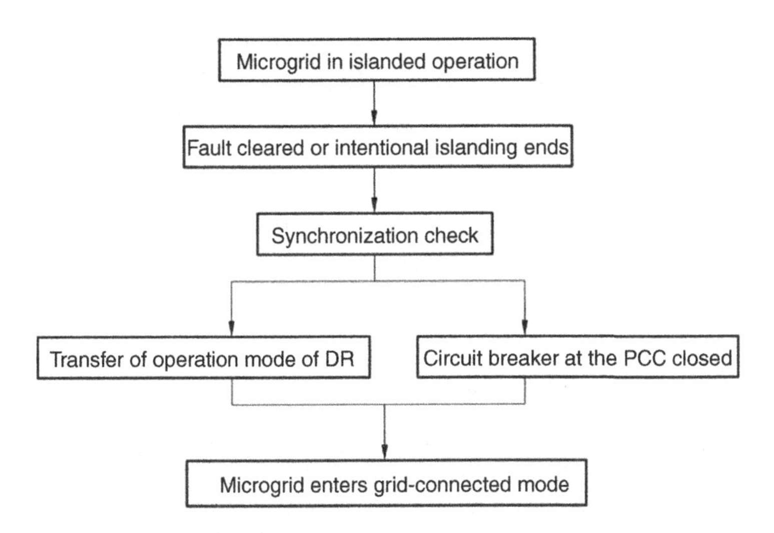

FIGURE 4.9 Flowchart of transfer from islanded mode to grid-connected mode.

After the microgrid is reconnected to the grid, the loads and DG previously taken out of the system are gradually reconnected to the system, and the microgrid enters the grid-connected mode. Figure 4.9 shows the flowchart of transfer from islanded mode to grid-connected mode.

4.3.2 Grid separation control

For transfer of a microgrid from grid-connected mode to islanded mode, quick and accurate islanding detection is a prerequisite. Currently, there are many islanding detection methods, from which a suitable one should be determined based on the specific conditions. Short-time unsmooth transfer or smooth transfer may be adopted, depending on whether the microgrid has loads that must be continuously powered.

4.3.2.1 *Islanding of microgrid*

The microgrid is a solution for the integration of DG to a distribution network. It changes the architecture of the distribution network and realizes bidirectional power flow. Traditionally, when the distribution system became unavailable, power supply to loads will be interrupted. With a microgrid, when the distribution system is out of service, the DG is expected to operate independently to power the loads. This is known as islanding, either intentional or unintentional, as shown in Figure 4.10. Intentional islanding means planned islanding with predetermined control strategies, and unintentional islanding means unplanned and uncontrollable islanding, which should be prevented.

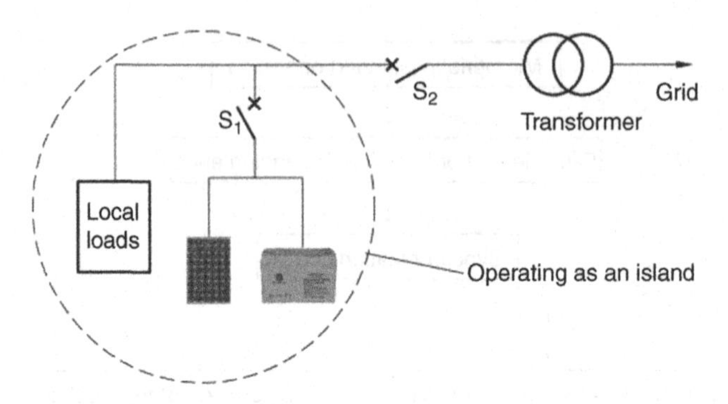

■ FIGURE 4.10 Schematic diagram of islanding.

Unintentional islanding does not meet the utilities' requirements on grid management. The status of an island system is unknown as it is not monitored by the utility and operates independently. Therefore, unintentional islanding is uncontrollable, poses high risks, and will cause the following problems:

1. Some lines that are considered to have been disconnected from all power sources may still be powered, endangering the maintenance personnel and consumers.
2. Normal closing of the grid is disturbed. When the DG in islanded operation is reconnected to the grid, the microgrid system may be asynchronous with the grid. In this case, when the circuit breaker is closed, it may be damaged and a high current may appear, thus damaging the DG, or even causing tripping of the grid.
3. The grid cannot control the voltage and frequency of the island system, thus causing damage to distribution equipment and loads. If, after islanding occurs, the DG cannot regulate its voltage and frequency and is not equipped with protection relay for limiting deviation, the voltage and frequency of DG will fluctuate remarkably, thus causing damage to distribution equipment and loads.

With the development of the microgrid and increasing penetration of DG, anti-islanding is essential. Anti-islanding is intended to prevent unintentional islanding. The key of anti-islanding is detection, which is the prerequisite for islanded operation of the microgrid.

4.3.2.2 Transfer from grid-connected mode to islanded mode

1. Unsmooth transfer
   Given the relatively long time taken for the low-voltage (LV) circuit breaker at the PCC to operate, power supply will be lost during this time during the transfer from grid connection to islanding. This is the so-called unsmooth transfer.

In the case of grid failure or outage, if voltage and frequency of the grid-tie bus are found beyond the normal range, or an intentional islanding order is sent from the energy management system, the connection/separation controller quickly disconnects the circuit breaker at the PCC and initiates transfer of control mode of the main power source after unimportant loads (or DG, as the case may be) are disconnected from the microgrid. The main power source switches to *U*/*f* control from *P*/*Q* control to realize constant-frequency and constant-voltage output and ensure voltage and frequency stability within the microgrid.

In this process, the islanding protection of DG acts and the DG quits service. After the main power source initiates islanded operation and resumes supply to important loads, the DG will automatically connect to the system and start operation. If all DGs are started at the same time, the island system will suffer a high impact. Therefore, they should be started separately. The energy management system should increase their output gradually. As the output is on the increase, the loads previously shed should be connected until the loads or DG outputs are beyond further regulation, and generation and consumption reach a new balance, thus realizing fast transfer from grid-connected mode to islanded mode. See Figure 4.11 for the flowchart of unsmooth transfer from grid-connected mode to islanded mode.

2. Smooth transfer

For a microgrid placing a higher requirement on reliability, smooth transfer is preferred. For this purpose, a high-power solid-state switch (with a closing or opening time less than 10 ms) is required to make up for the slow turn-on or turn-off of the mechanical circuit breaker, and the structure of the microgrid needs to be optimized.

As shown in Figure 4.12, important loads, a proper amount of DGs, and the main power source are connected to a section of the bus that is connected to the microgrid bus through a static switch, thus forming a subsystem that can maintain energy balance at the instant of isolation from the grid. Unimportant loads are directly connected to the main grid through the circuit breaker at the PCC.

In grid-connected mode, the microgrid often exchanges a large amount of power with the grid. This is especially the case for a low-capacity microgrid, which is mainly supported by the grid. When the microgrid is isolated, there will be a high power deficit, requiring the circuit with the solid-state switch to achieve balance between important loads and power output of DGs within a very short time after islanding occurs. In islanded mode, the ES or microturbines with self-regulation capability are used to maintain frequency and voltage stability within the microgrid; therefore, their capacities should be determined considering their output, demand of important loads, and maximum and minimum

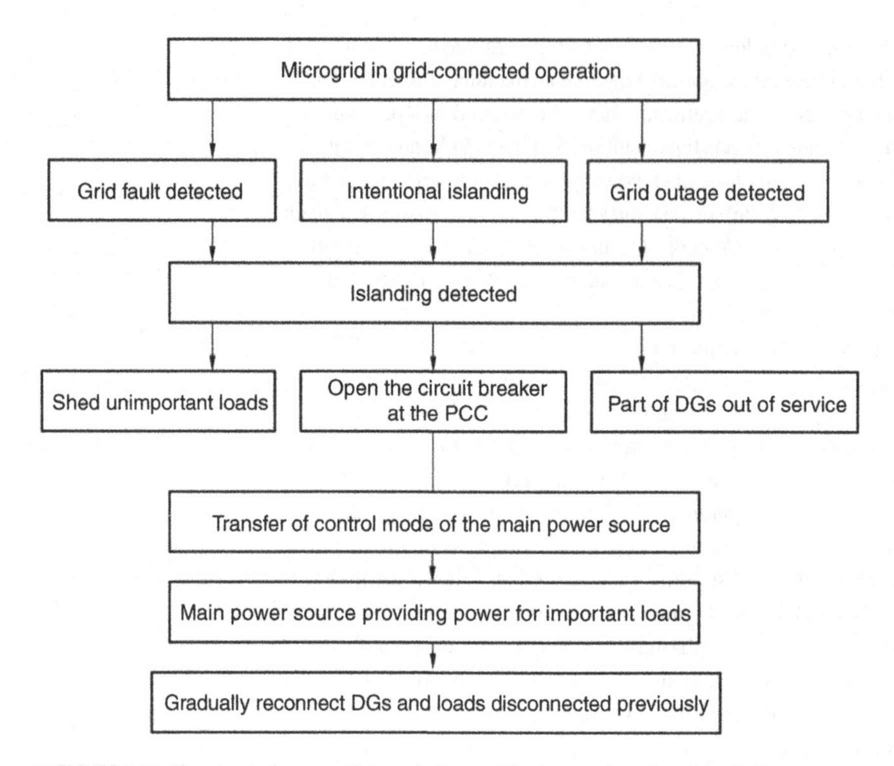

■ **FIGURE 4.11 Flowchart of unsmooth transfer from grid-connected mode to islanded mode.**

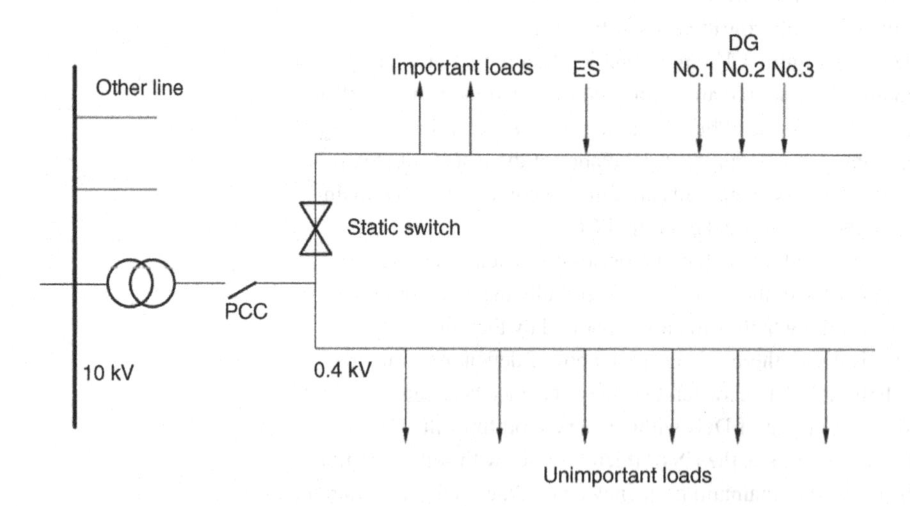

■ **FIGURE 4.12 Structure of microgrid using a solid-state switch.**

possible output of DG. A solid-state switch permits smooth transfer of the microgrid from grid-connected mode to islanded mode, and downsizes the service area of the microgrid in islanded mode.

In the case of grid failure or outage, if the system detects that the voltage or frequency of the grid-tie bus is beyond the normal range, or an intentional islanding order is sent from the energy management system, the grid connection/separation controller rapidly disconnects the circuit breaker at the PCC and the solid-state switch. As the switch can be quickly opened and closed, the main power source can start quickly and power important loads after the switch is opened, thereby ensuring continuous power supply to important loads. After the LV circuit breaker and circuit breaker for unimportant loads at the PCC are opened, the static switch is closed, and with large-capacity DGs restoring to operation, supply for unimportant loads is resumed gradually. See Figure 4.13 for the flowchart of smooth transfer from grid-connected mode to islanded mode.

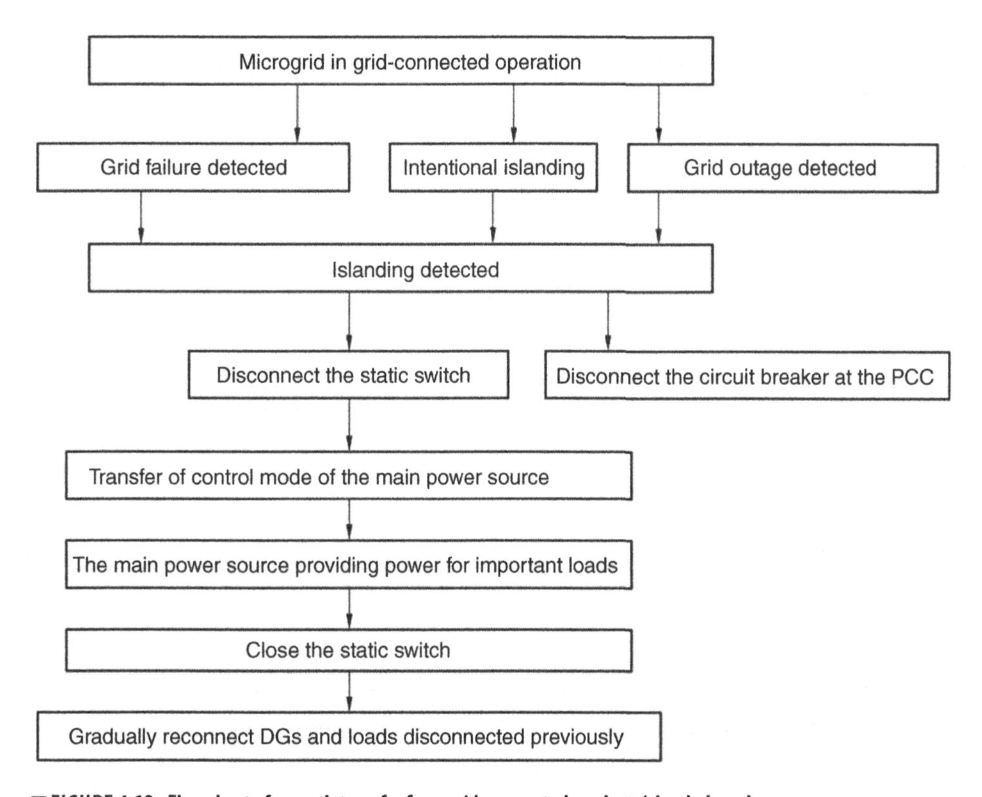

FIGURE 4.13 Flowchart of smooth transfer from grid-connected mode to islanded mode.

4.4 OPERATION

A microgrid may operate in grid-connected mode or islanded mode. Grid-connected operation means that the microgrid is connected to the distribution network at the PCC and exchanges power with it. When load is greater than the DG output, the microgrid absorbs power from the distribution network; otherwise, the microgrid injects the excess power to the distribution network.

4.4.1 Grid-connected operation

Grid-connected operation can realize economic and optimal dispatch, central dispatch by the distribution network, automatic voltage and reactive power control, and forecast of intermittent DG output, loads, and power exchange. See Figure 4.14 for the flowchart of grid-connected operation.

1. *Economic and optimal dispatch*: A microgrid in grid-connected operation is aimed at maximizing energy efficiency (i.e., to utilize

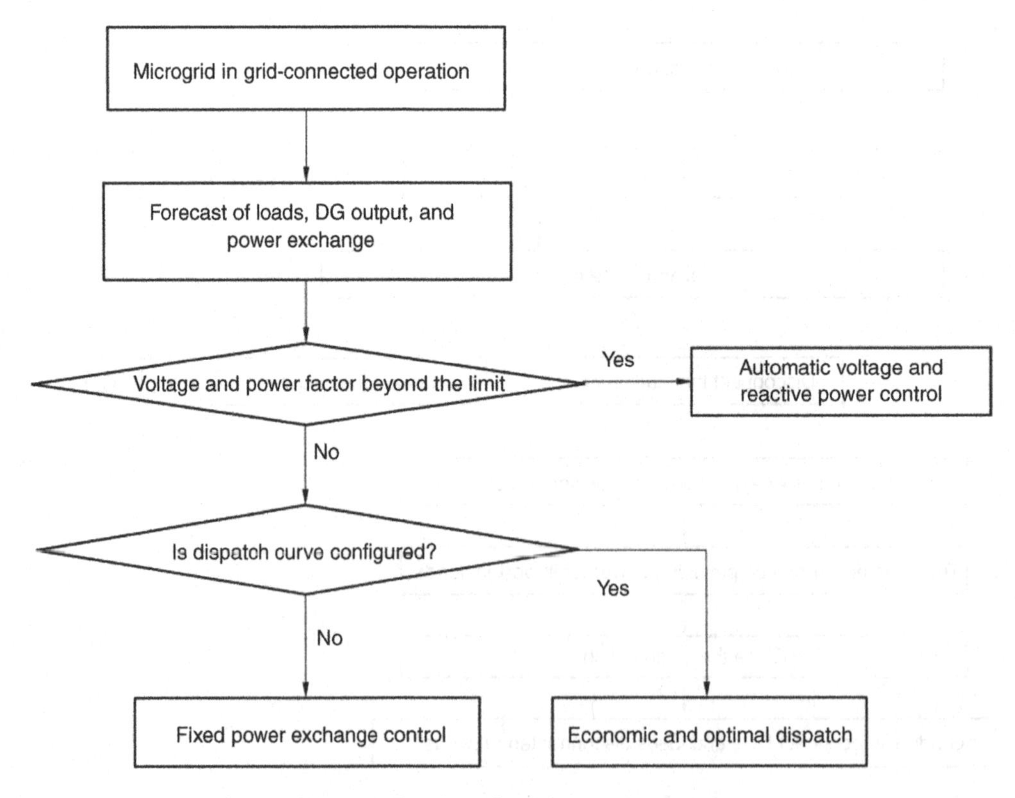

■ **FIGURE 4.14 Flowchart of grid-connected operation.**

renewable energy as practical as possible) across the system under the premise of security, and to shift loads by means of charge and discharge of ES and time-of-use tariff.

2. *Central dispatch by the distribution network*: The centralized control layer of microgrid interacts with the distribution network dispatch layer in real time, and sends the microgrid status (connected or disconnected) and amount of power exchange at the PCC to the dispatch center, and acts according to control over the microgrid status and power exchange setting by the dispatch center. When an order is received from the dispatch center, the centralized control layer regulates the DG output, ES, and loads to maintain balance of active power and reactive power. Central dispatch by the distribution network can be achieved by setting the power exchange curve either in the microgrid management system or according to the remote orders from the distribution network dispatch automation system.
3. *Automatic voltage and reactive power control*: The microgrid can be deemed as a controllable load with respect to the grid. In grid-connected mode, the microgrid is not allowed to participate in management of grid voltage, and it manages its power factor under a uniform power factor, and manages the bus voltage within the microgrid to some extent by regulating the reactive power compensation equipment and reactive output of DG.
4. *Forecast of intermittent DG output*: Short-time DG output can be forecast according to the weather information released by the meteorological bureau, historical meteorological information, and historical records of DG outputs.
5. *Load forecast*: The demand of various loads, including total loads, sensitive loads, controllable loads, and interruptible loads within a short period can be forecast according to historical records.
6. *Forecast of power exchange*: Power exchange through the common circuits can be forecast based on forecast of DG output and loads and preset charge and discharge curves of ES.

4.4.2 Islanded operation

Islanded operation is mainly intended for stability of the microgrid and for continuous power supply to as many loads after the microgrid is separated from the main grid. See Figure 4.15 for the flowchart of islanded operation.

1. *Load shedding in case of under-frequency or under-voltage*: Fluctuation of load or DG output beyond the compensation capacity of ES may result in a drop of system frequency and voltage. When they drop

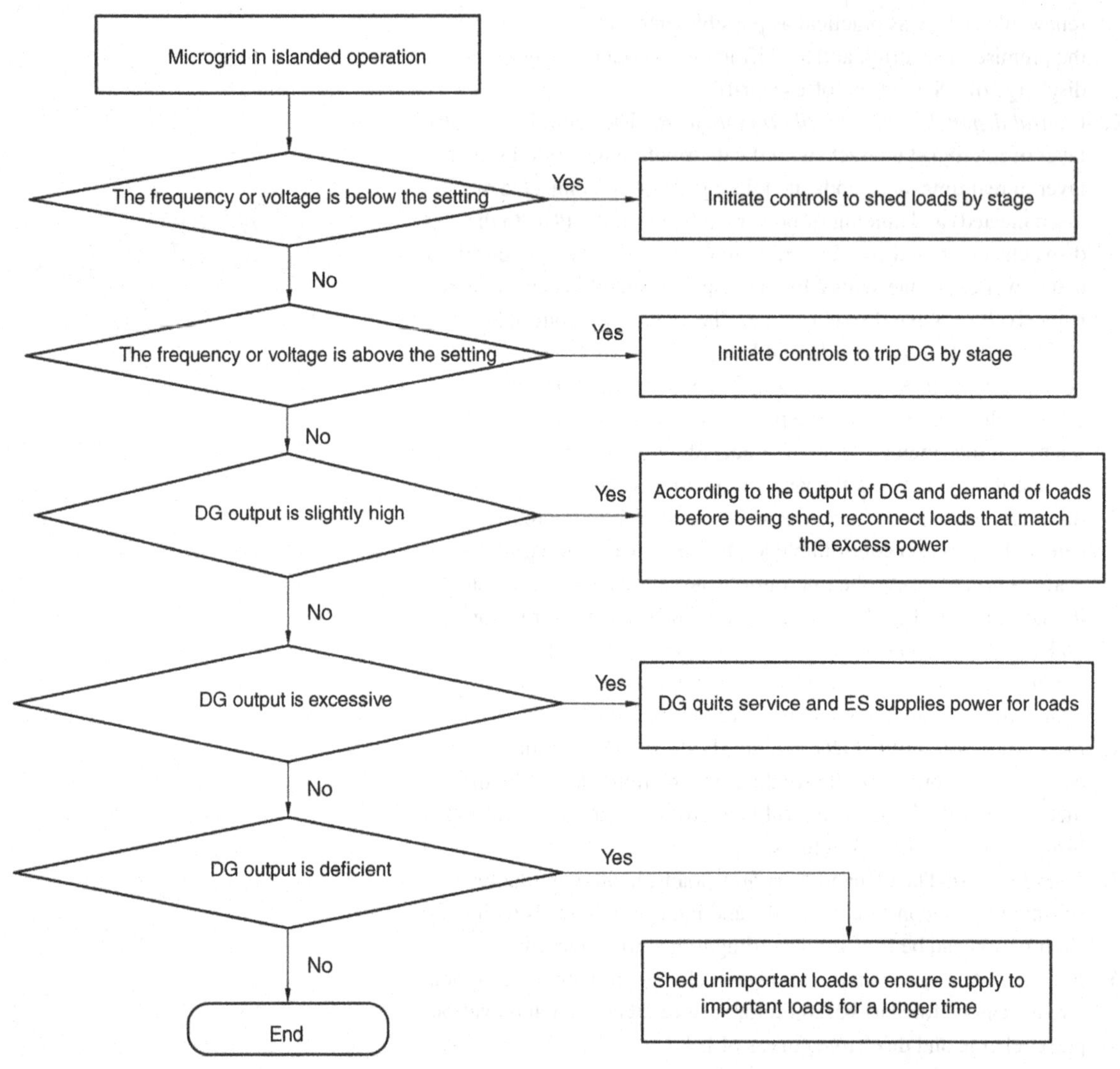

■ FIGURE 4.15 Flowchart of islanded operation.

below the settings, unimportant loads will be shed to prevent frequency collapse and voltage collapse.

2. *Generator tripping in case of over-frequency or over-voltage*: Fluctuation of load or DG output beyond the compensation capacity of the ES may result in a rise of system frequency and voltage. When they rise above the settings, the output of some DGs will be limited until the system frequency and voltage fall back to the normal range.
3. *Control in case of relatively high DG output*: In the case of high DG output, some loads previously shed can be reconnected to consume the excess power.

4. *Control in case of excessive DG output*: If DG output is excessive, and the system frequency and voltage are very high even though no loads are shed and all ESs are fully charged, the DG will quit service and loads are powered by the ES for a certain time, and then the DG is brought back to service.
5. *Control in case of deficient DG output*: If the dispatchable DG has been producing maximum output, and the SOC of ES is below the lower limit, unimportant loads will be shed to ensure continuous supply to important loads as long as possible.

Chapter 5

Protection of the microgrid

The same as a traditional electric power system, a microgrid is required to operate securely and stably and its relay protection should provide reliability, response speed, flexibility, and selectivity. When the microgrid is grid-connected, power can flow bidirectionally between the distribution network and the microgrid, different from the unidirectional flow found in traditional distribution networks. In addition, different from traditional rotating generators, the microgrid is flexibly connected to the grid based on power electronic technologies, which brings a difference to the relay protection.

5.1 SPECIAL FAULT CHARACTERISTICS OF DG

Different types of distributed generation (DG) sources are integrated to the grid in different modes:

1. *DC sources*: power sources producing direct current, including fuel cells, photovoltaic (PV) cells, and DC wind turbines. These sources are integrated to the grid via inverters, as shown in Figure 5.1a.
2. *AC–DC–AC sources*: power sources producing alternating current not at power frequency, including AC wind turbine and single-shaft microturbine. The alternating current needs to be converted to direct current and then back to alternating current before being delivered to the grid, as shown in Figure 5.1b.
3. *AC sources*: producing stable alternating current at power frequency, including asynchronous wind turbines and small synchronous generators. These sources are directly integrated to the grid without using any power electronic inverters, as shown in Figure 5.1c.

As stated, DG sources are connected to the grid either directly or via inverters, with the latter more often used.

Inverters are under either *P*/*Q* or *U*/*f* control. In grid-connected operation, all DGs are under *P*/*Q* control; in islanded operation, the master DG is under *U*/*f* control while the slave DGs are under *P*/*Q* control.

Microgrid Technology and Engineering Application

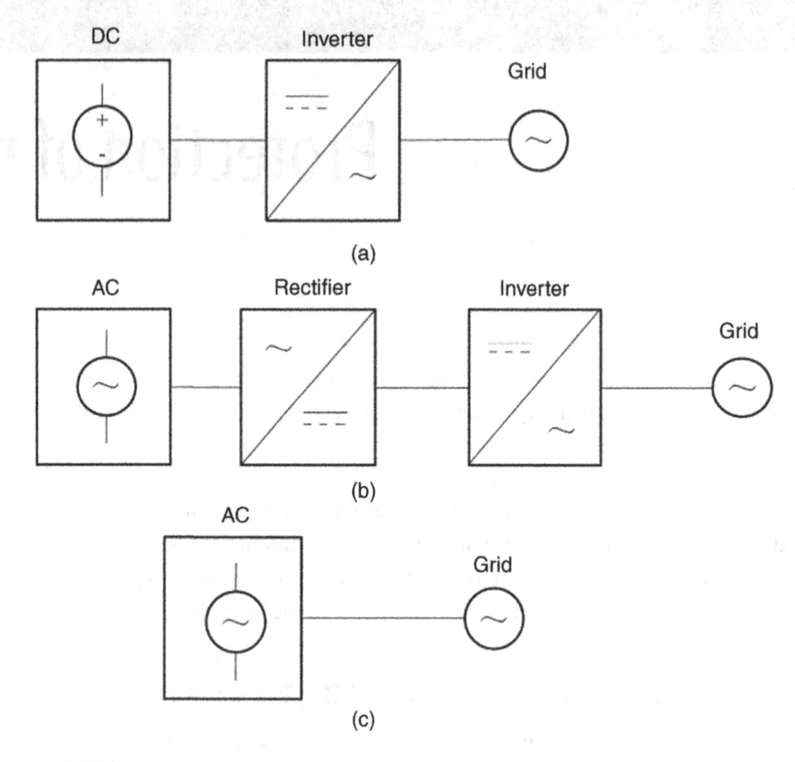

FIGURE 5.1 Schematic diagrams of DG sources connected to the grid. (a) DC sources, (b) AC–DC–AC sources, and (c) AC sources.

For inverters under *P*/*Q* control, the output current when a fault occurs should not be higher than $1.5I_n$ according to Q/GWD 147 – 2010 Technical Specifications for Interconnecting PV Stations with Grid and CGC/GF 001:2009 Technical Requirements and Test Methods for Inverters Exclusively for Integrating PV Power to the Grid. In the case of the three-phase short circuit, if the fault current is smaller than $1.5I_n$, the inverter is a constant-power source, the current rises, and the voltage drops; if the fault current is equal to $1.5I_n$, the inverter is a constant-current source, and after its protection acts, the inverter automatically quits service. In the case of asymmetric short circuit, the inverter is a constant-power positive sequence source, and the current rises; in the case of two-phase short circuit, the negative sequence voltage rises; and in the case of single-phase-to-earth fault, the zero sequence voltage rises.

For inverters under *U*/*f* control, in the case of the three-phase short circuit, the inverter is a constant-voltage constant-frequency source, and if the output power has not reached the maximum, the current rises and the output power increases; and if the output power has reached the maximum, the voltage drops and the undervoltage protection of the inverter acts; in the case of asymmetric short circuit, the inverter is a constant-power positive sequence source, and the current rises; in the case of two-phase short circuit, the negative sequence voltage rises; and in the case of single-phase-to-earth fault, the zero sequence voltage rises.

5.2 EFFECTS OF MICROGRID ON RELAY PROTECTION OF THE DISTRIBUTION NETWORK

5.2.1 Protection of a traditional distribution network

A traditional distribution network is a 10 kV network in one-way radial pattern or in hand-in-hand open-loop pattern.

Figure 5.2a shows a one-way radial distribution network, which is generally provided with traditional three-stage current protections, respectively current quick-break protection, specified time current quick-break protection, and specified time overcurrent protection. The protection is set as follows: current quick-break protection is set based on the maximum short-circuit current following a three-phase short-circuit fault at the end of the line and

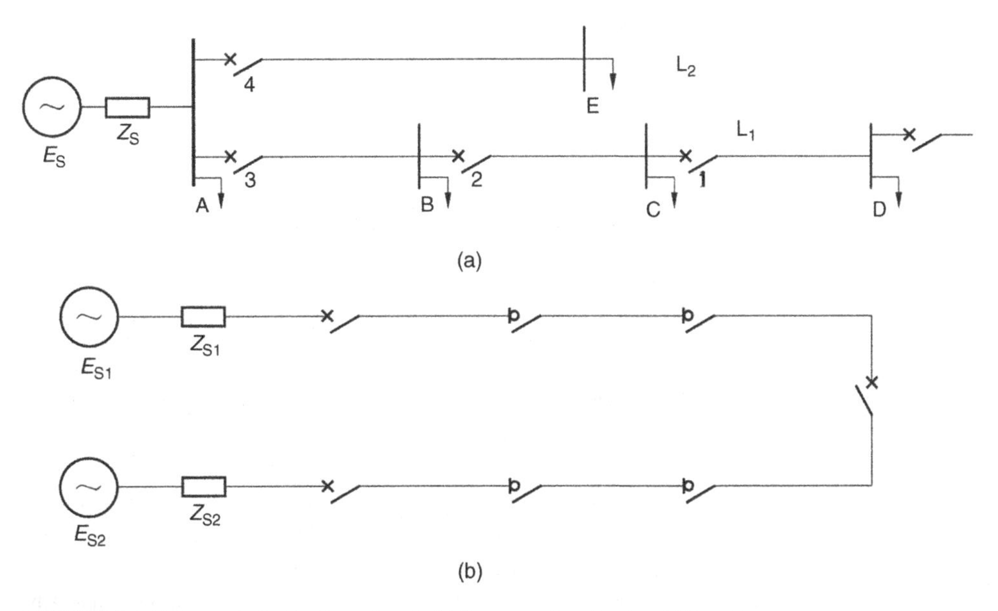

■ **FIGURE 5.2 Patterns of distribution network.** (a) One-way radial network and (b) hand-in-hand loop network.

cannot protect the entire line; specified time current quick-break protection is set based on the operating current of current quick-break protection for components of the neighboring line and can protect the entire line; specified time overcurrent protection is set based on the maximum load current of the line, serves as the backup for the protection of the neighboring line, and can protect the entire line. For nonterminal lines, three-stage current protection and other line protections are used together. The protection for terminal lines is simplified, generally including current quick-break protection (set to avoid the maximum three-phase short-circuit current appearing on the low-voltage (LV) side of the step-down transformer at the line end) and specified time overcurrent protection (set to avoid the maximum load current, with a time delay of 0.5 s). Cable lines mostly experience permanent faults and are not provided with reclosing function; while overhead lines are generally provided with this function.

Figure 5.2b shows a hand-in-hand loop network, in which the automatic circuit recloser realizes automatic distribution, and the stage current protection, automatic circuit recloser, and sectionalizer work together to isolate the faults. When a fault occurs on a line, the circuit breaker trips off and the voltage of the line is lost. When voltage loss is detected, all sectionalizers will open. When the circuit breaker is reclosed for the first time, the automatic circuit reclosers will switch in stage by stage according to the predefined time delays. When it comes to the fault section, the circuit breaker trips off again and the switches on both sides of the faulty section will detect fault voltage and block. Then, after the circuit breaker recloses again, the normal section will resume normal operation while the faulty section is isolated.

5.2.2 Protection of a traditional LV distribution network

An LV distribution system operates at 0.4 kV (380/220 V) and is usually provided with an LV circuit breaker (also called universal circuit breaker) with relay protection, fuse protection, and hot relay protection. According to GB 10963.1 – 2005/IEC 60898 – 1:2002 *Circuit-breakers for Overcurrent Protection for Household and Similar Installation-Part 1: Circuit-breakers for a.c. Operation*, there are three types of instantaneous current tripping: (1) the tripping range of type B is $3I_n$–$5I_n$, (2) type C $5I_n$–$10I_n$, and (3) type D $10I_n$–$20I_n$, and quick-break tripping time shall be less than 0.1 s. The time delay of action following a short-circuit has an inverse time feature, the conventional tripping current is $1.45I_n$, and the tripping time is less than 1 h ($I_n < 63$ A) or 2 h ($I_n > 63$ A).

5.2.3 Impacts of microgrid on relay protection of distribution network

A traditional distribution system is a radial one-end-source system, which requires no directional element for feeder protection, and is mostly provided with three-stage current protection. After a microgrid is connected to the distribution network, when a fault occurs, the DG sources in the microgrid, in addition to the system, also contribute to the fault current, forming a multi-source system and changing the short-circuit current level on a node. The type, location, and capacity of the DG sources affect the operation of relay protection of the distribution network.

The integration of a microgrid to a one-way radial distribution network mainly affects the distribution network in the following ways: reduction of sensitivity of infeed current protection for the line end; undesired operation of fault protection of neighboring lines; and reclosing failure. The following gives a detailed analysis on the impacts of the microgrid on relay protection of the distribution network when the microgrid is connected to the distribution network at different points.

1. *At the end of the feeder*: As shown in Figure 5.3, DG is integrated to the distribution network at the feeder end. When a fault occurs on point k_1 on the neighboring line L_2, the fault current may flow from the DG to k_1, causing the protections 1, 2, and 3 of L_1 to operate falsely.
2. *In the middle of the feeder*: As shown in Figure 5.4, the DG is integrated to the distribution network in the middle of the feeder. Similarly, when

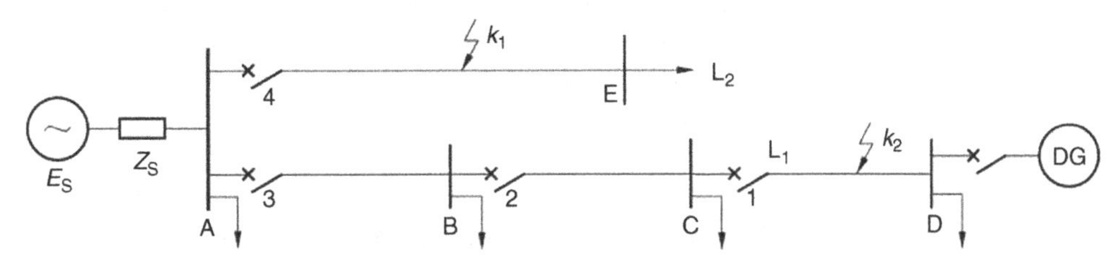

FIGURE 5.3 DG integrated to the distribution network at the feeder end.

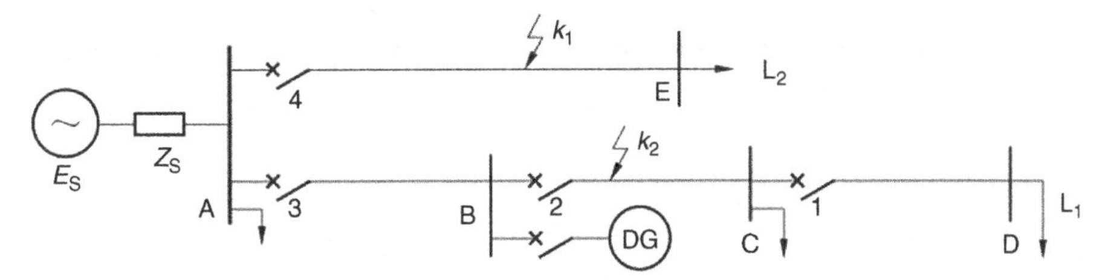

FIGURE 5.4 DG integrated to the distribution network in the middle of feeder.

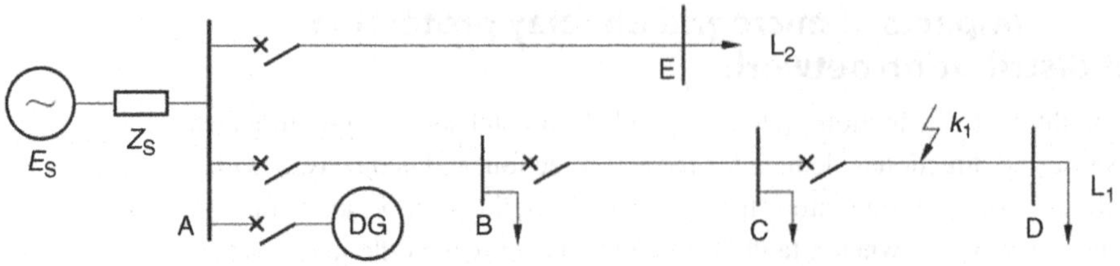

■ FIGURE 5.5 DG integrated to the distribution network at the head of feeder.

a fault occurs at point k_1 on the neighboring line L_2, the fault current may flow from DG to k_1, causing the protection 3 of L_1 to operate falsely; when a fault occurs at point k_2 on L_1, due to the infeed current produced by DG, the sensitivity of protection 3 is reduced, probably causing failure of operation, and therefore, the branch coefficient needs to be recalculated. With the DG connected to the distribution network, protection 2 needs to be reset based on the maximum operation mode. When protection 2 operates, the arc cannot be extinguished, causing reclosing failure of protection 2.

3. *At the head of the feeder*: As shown in Figure 5.5, the DG is integrated to the distribution network at the head of the feeder. In this case, all protections need to be reset based on the new maximum operation mode.

In a hand-in-hand loop distribution network, the automatic circuit recloser is used for automatic power distribution, and the stage current protection works together with the automatic circuit recloser and sectionalizer to isolate the faults. However, due to the presence of DG, when a fault occurs, the line still has a voltage, and thus the sectionalizer will not open, the fault cannot be isolated, and the backup power source cannot serve the normal section.

5.2.4 Impacts of a microgrid on protection of traditional LV distribution lines

Figure 5.6 shows the current path of load circuit in a microgrid system, in which $\dot{E}_S$ represents the potential of the microgrid system, $\dot{E}_{DG}$ the potential

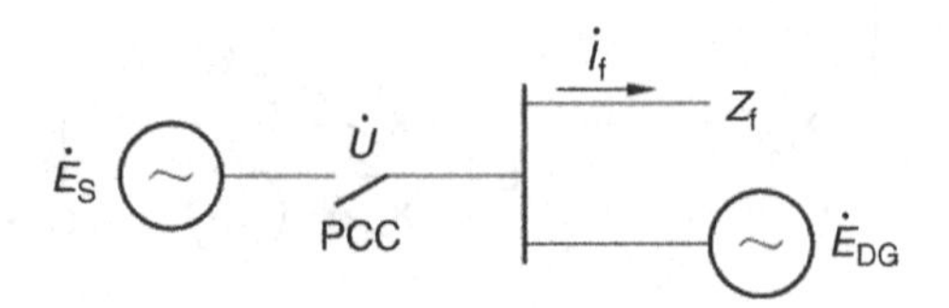

■ FIGURE 5.6 Path of load current in a microgrid system.

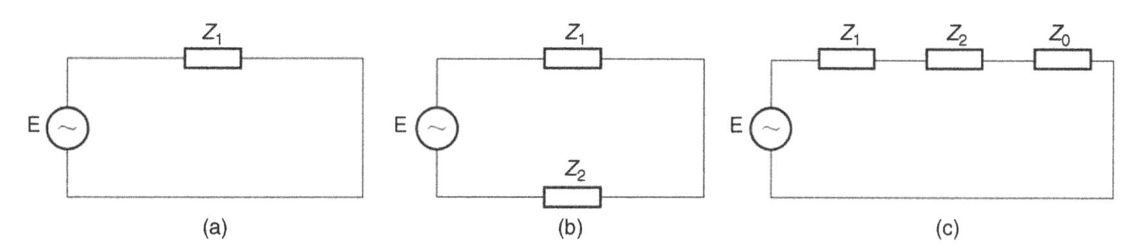

FIGURE 5.7 Complex sequence network diagram of various short circuits.
(a) Three-phase short circuit, (b) two-phase short circuit, and (c) single-phase short circuit.

of DG sources, PCC the point of common coupling, $\dot{U}$ the voltage at the PCC, Z_f loads of the microgrid, and $\dot{I}_f$ load currents. In normal grid-connected operation, $\dot{E}_S$ and $\dot{E}_{DG}$ jointly provide the load current; and in islanded operation, $\dot{E}_{DG}$ alone provides the load current.

Figure 5.7 gives a complex sequence network diagram for various faults based on the method of symmetrical components, where $\dot{E}$ represents the potential of the power source, and Z_0, Z_1, and Z_2 the zero-sequence impedance, positive-sequence impedance, and negative-sequence impedance, respectively. In grid-connected operation, $\dot{E}$ is equivalent to the sum of the grid source $\dot{E}_S$ and DG source $\dot{E}_{DG}$, and in islanded operation, $\dot{E}$ is the DG source $\dot{E}_{DG}$. Fault current analysis involves three-phase short circuit, two-phase short circuit, and single-phase-to-earth fault.

5.2.4.1 Three-phase short circuit

Figure 5.8 is a diagram of the three-phase short circuit. Taking short circuit in phase U as an example, it can be inferred from Figure 5.7a that

$$\left.\begin{aligned} \dot{I}_U &= \dot{I}_1 = \frac{\dot{E}_U}{Z_1} \\ Z_1 &= Z_{1S} + Z_{1L} \end{aligned}\right\} \tag{5.1}$$

where Z_{1S} is the positive sequence impedance of power source, and Z_{1L} the positive sequence impedance of line.

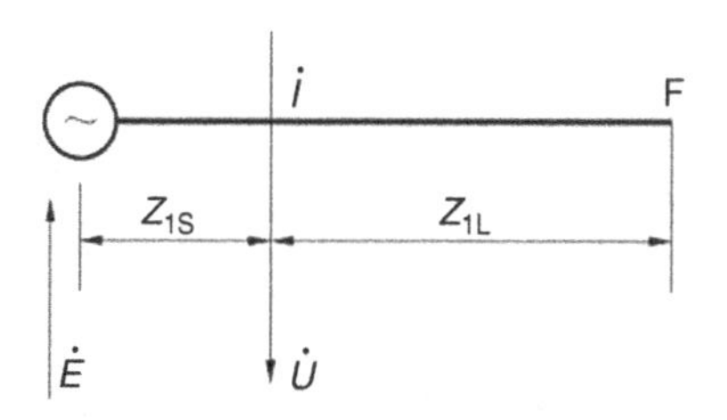

FIGURE 5.8 Diagram of three-phase short circuit.

5.2.4.2 Two-phase short circuit

Figure 5.9 is a schematic diagram of the two-phase short-circuit. Taking short circuit in phases V and W as an example, it can be inferred from Figure 5.7b that

$$\begin{aligned} &\dot{I}_1 = \dot{I}_2 = \dot{E}_U/(Z_1 + Z_2) \\ &\dot{I}_V = -\dot{I}_W = a^2\dot{I}_1 + a\dot{I}_2 = -j\sqrt{3}\dot{E}_U/(Z_1 + Z_2) \end{aligned}$$

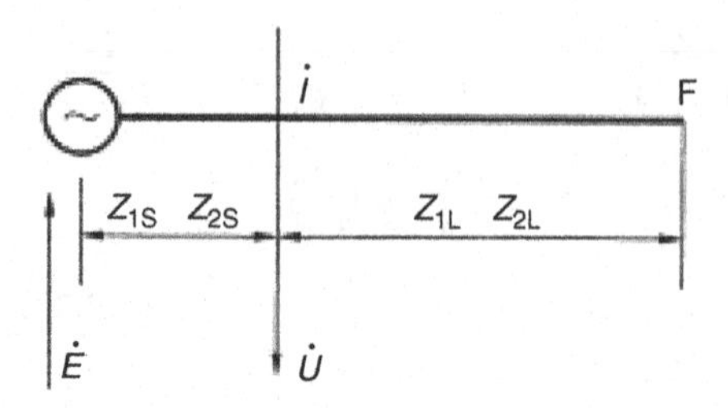

■ **FIGURE 5.9 Diagram of two-phase short circuit.**

where $Z_1 = Z_{1S} + Z_{1L}$, $Z_2 = Z_{2S} + Z_{2L}$

As Z_1 is approximately equal to Z_2, then

$$\dot{I}_V = -\dot{I}_W = -j\frac{\sqrt{3}}{2} \times \frac{\dot{E}_U}{Z_1} = -0.866j\frac{\dot{E}_U}{Z_1} \quad (5.2)$$

5.2.4.3 *Single-phase-to-earth fault*

Figure 5.10 is a diagram of single-phase-to-earth fault. Taking short circuit in phase U as an example, it can be inferred from Figure 5.7c that

$$\dot{I}_1 = \dot{I}_2 = \dot{I}_0 = \dot{E}_U/(Z_1 + Z_2 + Z_0) = \dot{I}_U/3$$

where $Z_1 = Z_{1S} + Z_{1L}$, $Z_2 = Z_{2S} + Z_{2L}$, and $Z_0 = Z_{0S} + Z_{0L}$, and Z_1 is approximately equal to Z_2, then

$$\dot{I}_U = \frac{3\dot{E}_U}{(2Z_1 + Z_0)} \quad (5.3)$$

For overhead lines, $Z_0 = 2Z_1$, then

$$\dot{I}_U = 0.75\frac{\dot{E}_U}{Z_1} \quad (5.4)$$

For cable lines, $Z_0 = 3.5Z_1$, then

$$\dot{I}_U = 0.55\frac{\dot{E}_U}{Z_1} \quad (5.5)$$

It can be seen that the three-phase short-circuit current is the greatest, followed by two-phase short-circuit current (0.87 times the three-phase short-circuit current) and single-phase-to-earth fault current (0.55 or 0.75 times the three-phase short-circuit current).

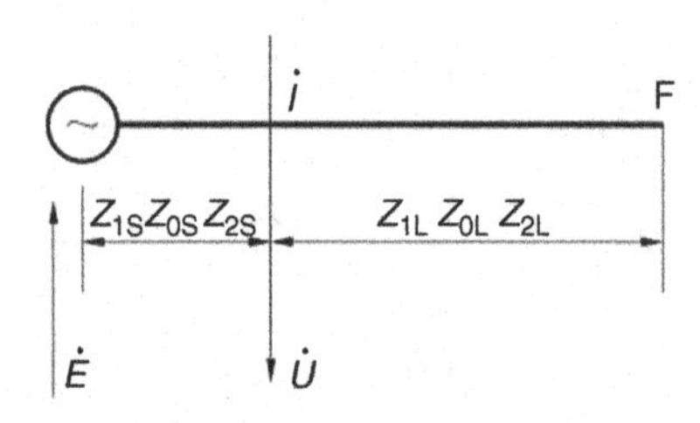

■ **FIGURE 5.10 Diagram of single-phase-to-earth fault.**

To carry out the above analysis and calculation, the impedance of the section from the location of protection to the equivalent power source, and the impedance of the section from the location of protection to the fault point should be identified. As the equivalent parameters of DG are hard to define, the short-circuit capacity method is used here to calculate the short-circuit currents in grid-connected operation and islanded operation.

In grid-connected operation, the grid source $\dot{E}_S$ operates in parallel with the DG source $\dot{E}_{DG}$, and the short-circuit capacity of the distribution transformer is

$$S_{1k} = \frac{S_{1N}}{U_{1k}\%} \tag{5.6}$$

where $U_{1k}\%$ is the short-circuit impedance (short-circuit voltage), and S_{1N} the capacity of the distribution transformer.

The short-circuit current provided by $\dot{E}_S$ is

$$I_{1k} = \frac{S_{1k}}{\sqrt{3}U} \tag{5.7}$$

Assuming that the overcurrent protection of inverters acts once the current rises above 1.5 times the rated current, the short-circuit capacity of $\dot{E}_{DG}$ is

$$S_{2k} = 1.5 S_{2N} \tag{5.8}$$

where S_{2N} is the capacity of $\dot{E}_{DG}$.

The short-circuit current provided by $\dot{E}_{DG}$ is

$$I_{2k} = \frac{S_{2k}}{\sqrt{3}U} \tag{5.9}$$

The maximum short-circuit current in grid-connected operation $I_k = I_{1k} + I_{2k}$.

In islanded operation, $\dot{E}_{DG}$ alone serves the loads, and the maximum short-circuit current I_k is equal to I_{2k}.

For instance, in Figure 5.6, $\dot{E}_S$ is a 10 kV distribution network connected with a distribution transformer with a capacity (S_{1N}) of 800 kVA, the short-circuit impedance $U_k\%$ is 4%, the voltage of LV bus, U is 400 V, and the capacity of the DG source $\dot{E}_{DG}$ (S_{2N}) is 600 kVA.

The maximum short-circuit current in grid-connected operation is

$$\begin{aligned} I_k &= I_{1k} + I_{2k} = \frac{800}{1.732 \times 0.04 \times 400} + \frac{1.5 \times 600}{1.732 \times 400} \\ &= 28.87 + 1.3 \approx 30.2\,(\text{kA}) \end{aligned}$$

The maximum short-circuit current in islanded operation is

$$I_k = I_{2k} = \frac{1.5 \times 600}{1.732 \times 400} \approx 1.3\,(\text{kA})$$

The short-circuit current in grid-connected operation, mainly provided by the power source of the distribution network, is much greater than that in islanded operation. In grid-connected operation, when a fault occurs on the LV distribution network, the system source and DG source contribute to the fault current, and therefore, the fault current is greater than that produced solely by the system source. In islanded operation, when a fault occurs on the LV distribution network, the DG alone provides the fault current, which is relatively small. In addition, in islanded operation, the output current of inverters is limited within $1.5I_n$ following a fault on the distribution network, the tripping time of a traditional LV circuit breaker with relay protection is nearly 1 h ($I_n < 63$ A), and thus the fault cannot be isolated rapidly.

5.3 MICROGRID OPERATION AND PROTECTION STRATEGIES

Microgrid protection is required to deal with the influence caused on the traditional distribution system by integrating the microgrid and meeting the requirement for protection brought by islanded operation. The integration of multiple DGs and energy storage (ES) changes the fault characteristics of the distribution system, and makes the variation of electrical variables very complicated in the case of a fault. The traditional protection principles and fault detection methods may fail to correctly locate the fault. Proper operation and protection strategies are the key to reliable operation of the DG system. As the microgrid is intended for grid-connected operation and islanded operation, its protection and control are very complicated. According to current practices, control and protection are a core microgrid enabling technology.

Before the concept of microgrid was introduced, the DGs integrated to the distribution network were not allowed to operate independently. Hence, besides the basic protection functions, the inverters should be able to prevent the occurrence of islanding, and automatically take the DGs out of service following a fault. The main protection strategies are as follows:

1. The DG is automatically taken out of service following a fault on the distribution network and the protection of the distribution network is not affected.

2. The capacity and PCC of DG are limited and the distribution network remains unchanged.
3. Measures for limiting the fault current are taken, for example, a fault current limiter is used to minimize the effects of DG after a fault occurs, and the distribution network remains unchanged.

A microgrid should be able to operate in parallel with the grid and in islanded mode. It shall meet the following basic requirements:

1. In grid-connected operation, if a fault occurs on the microgrid, the microgrid protection should operate reliably to clear the fault. For example, when the electrical equipment in the LV distribution network fails, the protection of the distribution network should operate to remove the equipment to ensure secure and stable operation of the microgrid.
2. When an instantaneous fault occurs on the distribution network, the protection of the distribution network should operate quickly to clear the fault to maintain uninterrupted operation of the microgrid.
3. In the case of power loss of the distribution network, the islanding protection of the microgrid operates to isolate the microgrid from the distribution network and the microgrid switches to islanded operation.
4. In islanded operation, when a fault occurs on the microgrid, the protection should operate reliably to clear the fault to ensure secure and stable operation of the microgrid;
5. Upon recovery of the distribution network, the microgrid is reconnected to the grid.

5.4 PROTECTION SCHEME FOR DISTRIBUTION NETWORK CONNECTED WITH A MICROGRID

5.4.1 Requirements of microgrid on primary equipment and relay protection of distribution network

In a traditional one-way radial distribution network, a circuit breaker is provided only at the power source end. While in a traditional hand-in-hand loop network, circuit breakers are provided at the power source end and the open-loop points, and sectionalizers are provided at other points. Since the sectionalizers are incapable of isolating faults, repeated reclosing operations are needed to isolate the fault. One purpose of the microgrid is to improve power reliability and quality, which requires quick fault isolation.

With the integrated microgrid, the traditional primary equipment in the distribution system cannot realize quick fault isolation, and therefore the following configuration is required:

1. Distribution networks at 10 kV or above should all be provided with a circuit breaker.
2. 0.4 kV LV distribution networks should be provided with a miniature circuit breaker supporting remote control.
3. The integration of the microgrid should not cause any change to the earthing pattern of the 0.4 kV LV distribution network.
4. In islanded operation, earthing of DGs should be considered.

5.4.2 Differential relay protection of the distribution network

To solve the problems caused by integration of the microgrid, the mature differential protection scheme for high-voltage (HV) systems is configured for the distribution network as the main protection and simple overcurrent protection is configured as backup. Theoretically, Kirchhoff's law-based differential protection is the best scheme for the buses, lines, and transformers in substations. For this protection, only the currents on both sides of the protected object are collected and compared based on simple criteria and the protected object has a high sensitivity.

5.4.2.1 Main protection – differential protection

Figure 5.11 presents the protection for a distribution network containing multiple microgrids. A 10 kV distribution network can be divided into multiple areas based on the protected object, respectively, line differential protection area, bus differential protection area, and distribution transformer differential protection area, as shown in the figure. For the distribution transformer, the differential current should be calculated taking into account the effect of the $Y \rightarrow \Delta$ or $\Delta \rightarrow Y$ current compensation method. For other protection areas, if the forward current direction is presumed to be from bus to line, the differential current is the vector sum of currents on both sides.

In differential protection, differential protection actuating criteria and percentage restraining criteria form an AND gate, as shown in Figure 5.12.

The differential protection actuating criteria are

$$I_{\mathrm{d}} \geq I_{\mathrm{OP}_0} \tag{5.10}$$

The percentage restraining criteria are

$$I_{\mathrm{d}} > kI_{\mathrm{r}} \tag{5.11}$$

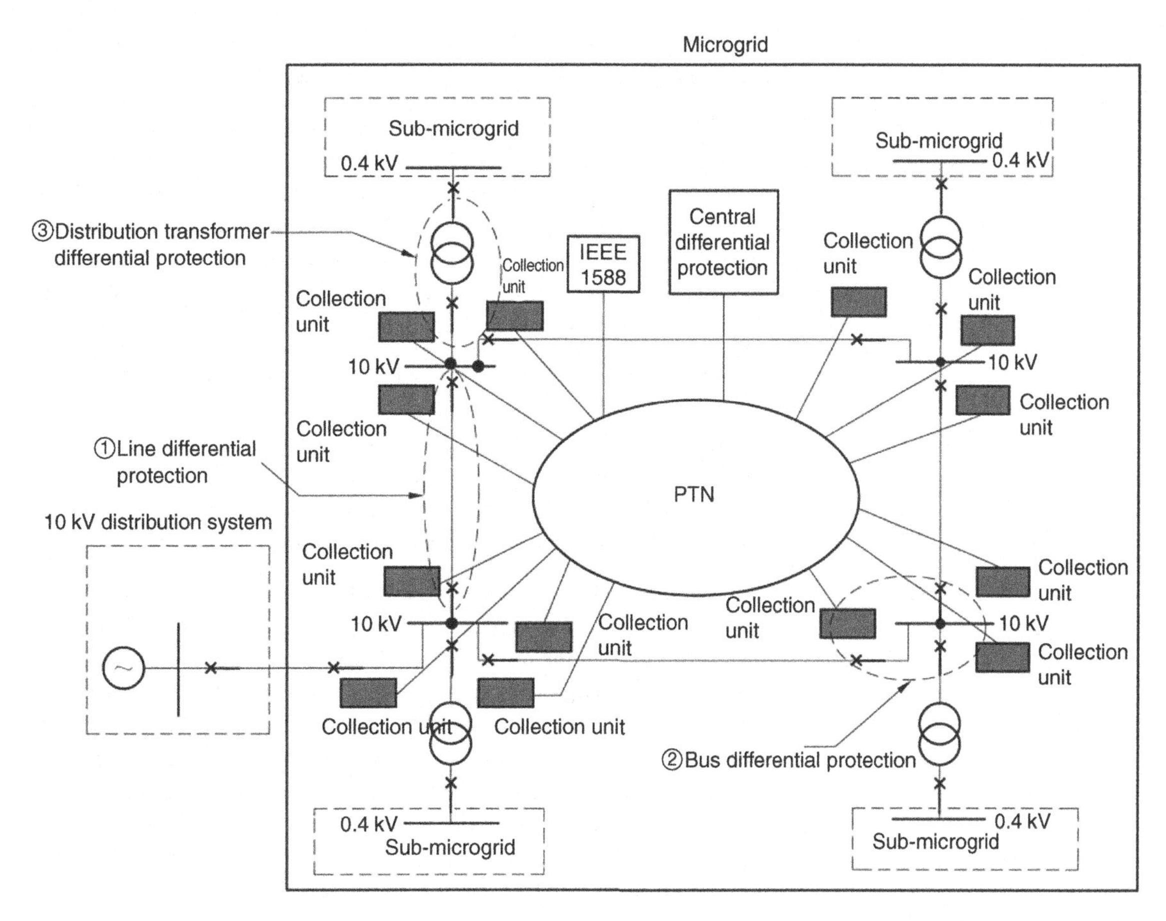

■ FIGURE 5.11 Schematic diagram of protection for a distribution system containing multiple microgrids.

where, I_d means the differential current, $I_d = |\dot{I}_1 + \dot{I}_2 + \cdots + \dot{I}_k|$; I_{OP_0} the setting value of the actuating criteria; I_r the restraining current, $I_r = |\dot{I}_1| + |\dot{I}_2| + \cdots + |\dot{I}_k|$; k the percentage restraining coefficient; and $\dot{I}_k$ currents on both sides of the protected object.

The smart collection unit provided for the circuit breaker interacts with the differential protection through the Packet Transport Network (PTN) with the sampled value and Generic Object Oriented Substation Event (GOOSE)

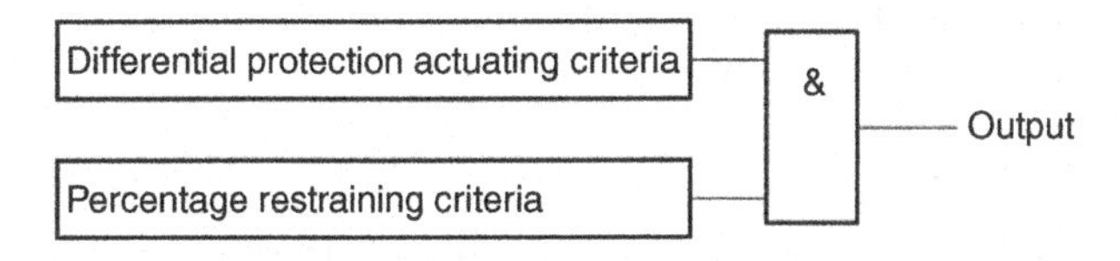

■ FIGURE 5.12 Block diagram of current differential protection.

messages specified in IEC 61850. The IEEE 1588 clock synchronization protocol can realize a synchronization precision of 100 ns.

The IEC 61850 is the most complete standard on substation automation to date, and has been widely applied in smart substations. It defines GOOSE and switching Ethernet with priority and virtual local area network flag (IEEE 802·1Q), which ensure real-time transfer of messages. Using the multicast application association model based on the publisher/subscriber communication principles, the GOOSE communication mechanism is an effective solution for real-time data transfer from one data source to multiple receivers.

The IEEE 1588, a precision clock synchronization protocol for distributed measurement and control systems, was released in 2002 and has a precision of μs. In 2008, IEEE 1588 V2 was released, which further specified and optimized the precision, security, and redundancy. In the second and third revisions of IEC 61850, experts of the working group proposed to apply the IEEE 1588 protocol in the substation-wide automation system.

PTN is a new optical transmission network architecture. It provides a layer between IP services and optical transmission media at the bottom layer, and is designed to meet the requirements of sudden increase of packet services and statistical multiplexing. Although mainly intended for packet services, it also supports other service types. It helps reduce the overall cost while inheriting such traditional advantages of optical transmission as high availability and reliability, effective bandwidth management and traffic engineering, convenient operation administration and maintenance, scalability, and high safety. At present, clock synchronization based on IEEE 1588 V2 has been widely applied in PTN.

Differential protection is based on the three-layered structure described in Chapter 2. Figure 5.13 presents a diagram of the differential protection system based on three-layer control areas, respectively, the smart collection unit at the local control layer, differential protection at the centralized control layer, and distribution network dispatch system at the distribution network dispatch layer, consistent with the three-layer architecture of the microgrid control system. It best suits protection against and control of transient faults of independent microgrids. See Figure 4.1 "Schematic diagram of three-state control system of an independent microgrid" in Chapter 4. For reliability purposes, the differential protection at the central control layer is configured in duplicate.

The smart collection units at the local control layer have the following functions:

1. Collect the voltage and current at the installation points and switch position;

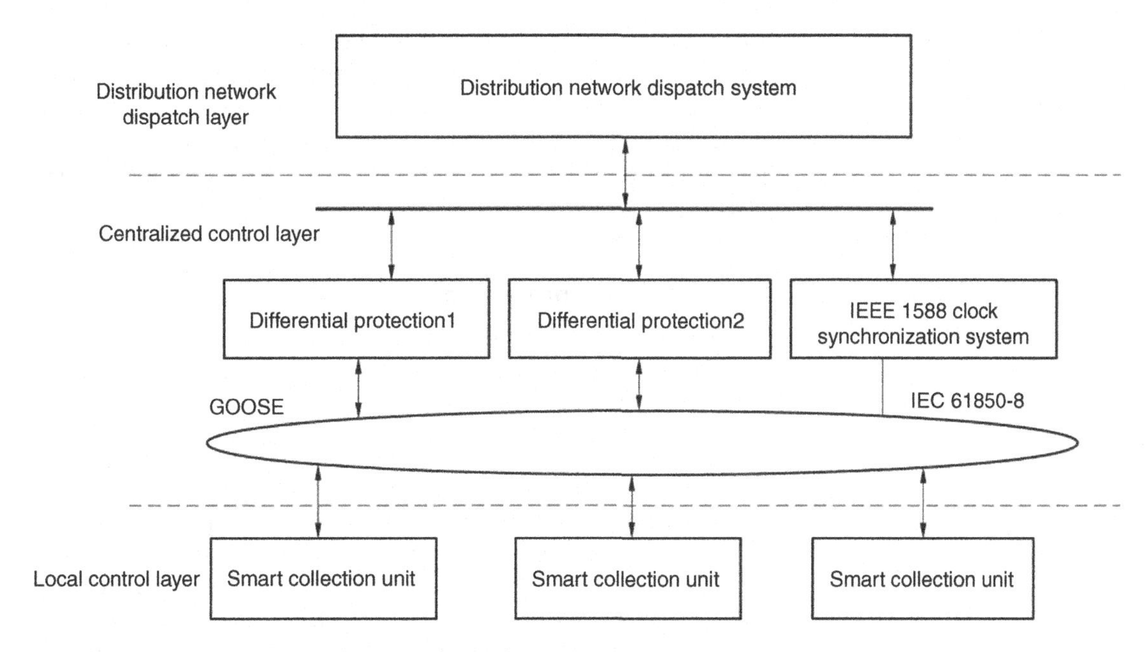

FIGURE 5.13 Schematic diagram of central differential protection system based on three-layer control architecture.

2. Receive and respond to the tripping and reclosing orders from the differential protection at the centralized control layer;
3. Provide back-up protection for equipment at the local control layer;
4. Send fault information and other operation information;
5. Provide back-up protection.

The differential protection at the centralized control layer has the following functions:

1. Receive sampled current values and status information sent from the smart collection units;
2. Determine the differential protection criteria based on sampled current values;
3. Provide circuit breaker failure protection;
4. Identify the faulty area and give tripping orders;
5. Send fault information to the distribution network dispatch layer.

The differential protection at the centralized control layer collects current and status information from all nodes of the distribution network. This is in nature a network-based differential protection and realizes quick and automatic fault location and isolation.

Differential protection is configured in duplicate, to prevent loss of protection of the entire system when the differential protection is out of service due to hardware failure. The two protection systems are both in service under normal conditions. When one of them fails and quits service, its logic judgment and tripping output are blocked, while the remaining one is not affected.

5.4.1.2 Back-up protection

In differential protection, circuit breaker failure protection is provided to make the neighboring circuit breaker operate and isolate the fault when a circuit breaker fails.

Distribution networks at 10 kV and above are provided with dual differential protection systems, ensuring high reliability, response speed, sensitivity, and selectivity. In addition to the highly reliable main protection, simple back-up protection, such as specific time-delay overcurrent or distance protection and directional overcurrent protection, may also be provided by making use of the smart collection units to prevent complete loss of protection of the distribution network when the central differential protection fails and quits service.

Local smart collection units are provided with back-up protection. Specifically, the local line collection unit is provided with distance or overcurrent protection as the back-up protection for lines and buses; the local transformer collection unit is provided with overcurrent protection as the back-up protection for transformers; the HV side of the distribution step-up transformer is provided with specific time-delay directional overcurrent protection (as illustrated in Figure 5.14a) as the back-up protection against internal faults of the transformer. It can also serve as the back-up protection for internal faults of the submicrogrids by setting it to prevent overcurrent flowing to the power source and be sensible to some extent to 0.4 kV bus faults. The distribution step-down transformer is configured with specific time-delay overcurrent protection as shown in Figure 5.14b, which is set based on the maximum load current.

5.4.3 Protection of LV distribution network based on directional impedance relay

In the LV distribution network containing a microgrid, forward and reverse impedance relays are provided for feeder units with DGs as distance protection, in which the forward impedance relay operates without time delay and is used for protection of the outgoing line. The reverse impedance

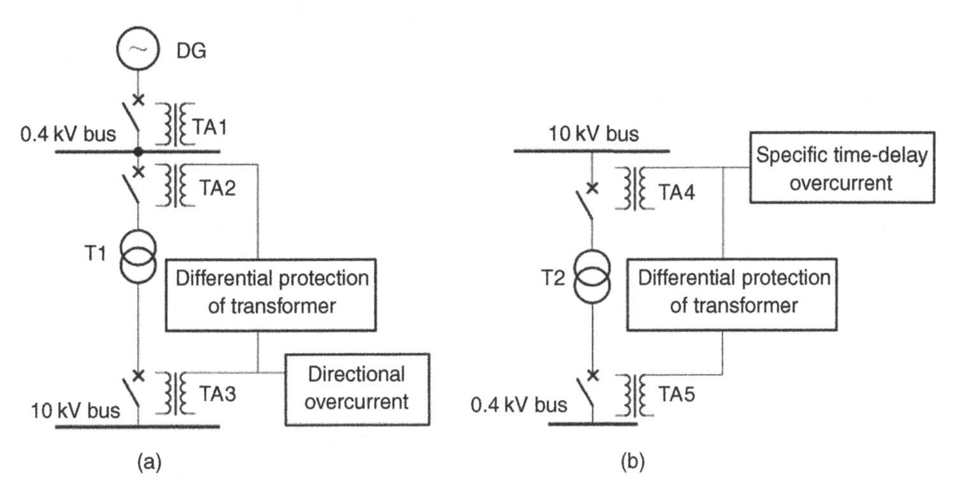

FIGURE 5.14 Back-up protection for distribution transformers. (a) Distribution step-up transformer and (b) distribution step-down transformer.

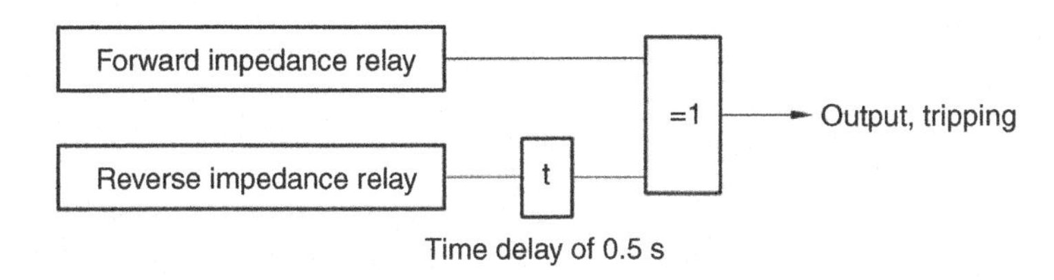

FIGURE 5.15 Feeder unit with DG.

relay operates with a time delay of 0.5 s and is used for protection of the LV bus (see Figure 5.15). Other feeder units without DG (load lines) are provided with forward impedance relay as distance protection, which operates without time delay and is used for protection of the outgoing line (see Figure 5.16).

The impedance relays use an overcurrent-actuating member, are polarized with memory positive sequence voltage, and have biasing impedance characteristics considering the small angle of impedance for a 0.4 kV voltage class.

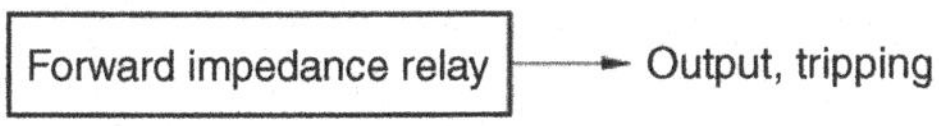

Forward impedance relay → Output, tripping

FIGURE 5.16 Feeder unit without DG.

Criteria of current-actuated element

$$I > I_{set} \tag{5.12}$$

Criteria of the forward impedance relay

$$90° + \theta < \arg\frac{\dot{U}_{|0|}}{\dot{U} - Z_{set}\dot{I}} < 270° + \theta \tag{5.13}$$

Criteria of the reverse impedance relay

$$-90° + \theta < \arg\frac{\dot{U}_{|0|}}{\dot{U} - Z_{set}\dot{I}} < 90° + \theta \tag{5.14}$$

where, for phase-to-phase impedance relay, $\dot{U} = \dot{U}_{\Phi\Phi}, \dot{I} = \dot{I}_{\Phi\Phi}$ $(\Phi\Phi = \mathrm{UV, VW, WU})$ for phase-to-ground impedance relay, $\dot{U} = \dot{U}_{\Phi}, \dot{I} = \dot{I}_{\Phi} + K\dot{I}_0[\Phi = \mathrm{U, V, W}, K = (Z_0 - Z_1)/Z_1]$; z_{set} is the setting value of impedance, and θ is the deviation angle of impedance, which is −30°.

The forward impedance relay is set to avoid the maximum load. When phase-to-phase or phase-to-ground short circuit occurs on the line, it operates without time delay to trip the circuit breaker of this line.

The reverse impedance relay should be set to avoid the short-circuit current of the bus or the setting value of the current quick-break protection for the line end on the HV side of the transformer. It covers the transformer and sometimes part of the HV line and LV line. Its setting value is relatively small so as to avoid unintentional operation when short circuit occurs on the HV side. In the case of phase-to-phase or phase-to-ground short circuit on the LV bus, the reverse impedance relay of the feeder unit with DG operates with a time delay of 0.5 s to trip the circuit breaker of this line and disconnect the DG from loads. In the case of short circuit on the LV line, transformer or HV line, although the reverse impedance relay of this feeder unit operates, the 0.5 s time delay ensures no protection output and no tripping of the circuit breaker of this line; and when the fault is cleared, the relay is automatically recovered.

This distance protection is in service in islanded operation and in grid-connected operation. In islanded operation (the microgrid is separated from the grid at the PCC), it is the sole protection for the microgrid. In grid-connected operation, it can also trip the circuit breaker of the 0.4 kV LV distribution system.

Figure 5.17 shows an analysis on the operation of microgrid protections following a fault at different points in islanded operation. When a fault occurs

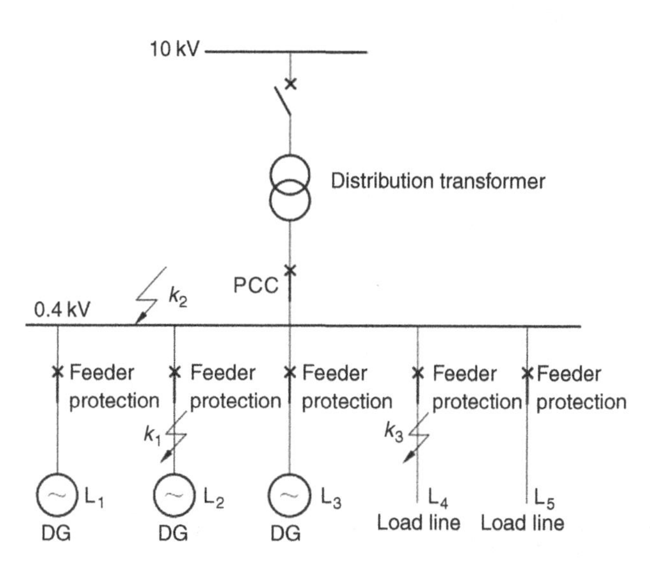

FIGURE 5.17 Schematic diagram of protection for microgrid in islanded operation.

at point k_1, the forward impedance relay of L_2 will operate; the forward impedance relays of L_1 and L_3 do not operate while the reverse impedance relays operate; as the current through L_4 and L_5 is approximately zero, neither the forward nor reverse impedance relays operate. After the fault is cleared by the protection of L_2, the reverse impedance relays of L_1 and L_3 are reset.

When a fault occurs at point k_2, the forward impedance relays of L_1, L_2, and L_3 do not operate while the reverse impedance relays operate with a time delay to trip the circuit breaker; neither the forward nor reverse impedance relays of L_4 and L_5 operate.

When a fault occurs at point k_3, the forward impedance relay of L_4, and reverse impedance relays of L_1, L_2, and L_3 operate. After the fault is cleared, the reverse impedance relays of L_1, L_2, and L_3 are recovered. Neither the forward nor reverse impedance relay of L_5 operates. The forward impedance relay of L_4 operates to trip the circuit breaker and clear the fault.

The protections operate in such a process: after the current element is actuated, the forward impedance relay operates to trip the circuit breaker on the branch; the reverse impedance relay operates with a time delay (to ensure the fault on the feeder is cleared) to trip the circuit breaker on this branch.

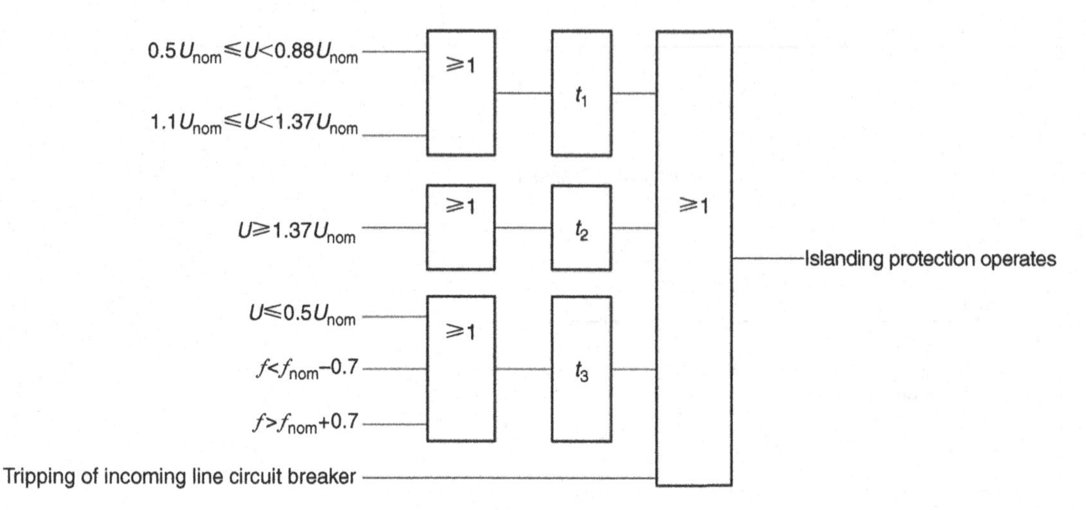

■ **FIGURE 5.18 Logical diagram of islanding protection at the PCC.**

5.4.4 Protection at the PCC

Islanding protection, overcurrent protection, and recloser for synchronous grid connection are the main protections provided at the PCC. Islanding protection operates to quickly trip the circuit breaker at the PCC when power loss of the distribution network is detected in grid-connected operation and thereafter, the microgrid switches to islanded operation. Overcurrent protection operates to trip the circuit breaker at the PCC in the case of a fault on the LV incoming line or LV bus in grid-connected operation. The recloser for synchronous grid connection can automatically connect the islanded microgrid to the distribution network when the distribution network restores to normal operation.

For islanding protection purposes, electrical variable-based passive detection and communication-based active detection are combined. As shown in Figure 5.18, the time of electrical variable detection is limited as required in IEEE std. 929, that is, $t_1 = 2$ s, $t_2 = 0.04$ s, and $t_3 = 0.1$ s. Communication-based active islanding detection is reliable and easy to achieve, in which the smart collection unit sends the status of each circuit breaker to the central control and protection through the network and the central control and protection conducts islanding detection. When power loss of the grid is detected, the islanding protection operates.

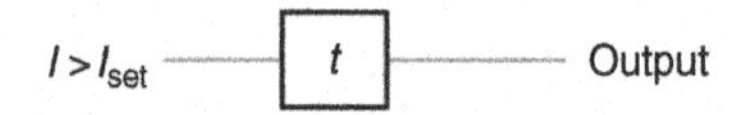

■ **FIGURE 5.19 Logical diagram of overcurrent protection.**

Figure 5.19 shows the logical diagram of overcurrent protection. Overcurrent protection operates to trip the circuit breaker at the PCC following a fault on the LV incoming line or LV bus.

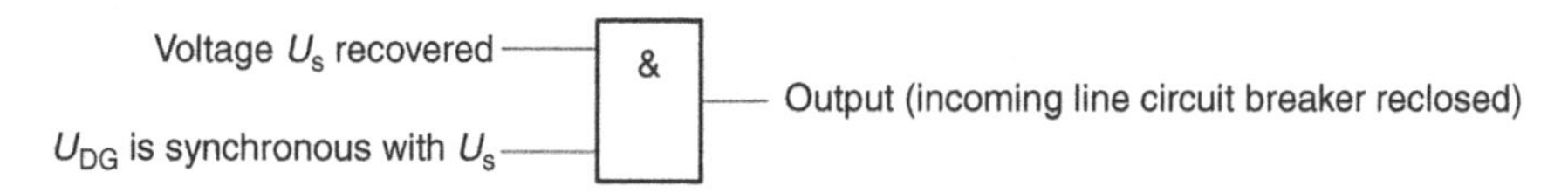

FIGURE 5.20 Logical diagram of synchronous grid connection.

Figure 5.20 shows the logical diagram of the recloser for synchronous grid connection. When it is detected that the distribution network restores to normal operation, and the bus voltage of the microgrid in islanded operation is synchronous with the voltage of the distribution network, the recloser for the synchronous grid connection operates to connect the microgrid to the grid.

Chapter 6

Monitoring and energy management of the microgrid

The monitoring and energy management system of the microgrid serves real-time, extensive monitoring of distributed generation (DG), energy storage (ES), and loads within the microgrid. In grid-connected operation, islanded operation, and during transition between operation modes, it controls and optimizes the DG, ES, and loads, thereby ensuring secure and stable operation of the microgrid at the maximum energy efficiency.

6.1 MONITORING

6.1.1 Structure of the monitoring system

The monitoring system coordinates with the local control and protection and remote distribution dispatch, and has the following functions:

1. Real-time monitoring of supervisory control, data acquisition, and DG;
2. Service management: forecast of power flow (including tie line power flow, DG node power flow, and load flow) and DG output; DG output control and power balance control;
3. Smart analysis and decision-making: optimized dispatch of energy.

Figure 6.1 shows the structure of the monitoring system.

By collecting information of DGs, lines, the distribution network, and loads in real time, the monitoring system monitors the power flow across the microgrid and adjusts the operation of the microgrid in real time based on operation constraints and energy balance constraints. In the monitoring system, energy management is the core that integrates the DG, load, ES, and the point of common coupling (PCC). Figure 6.2 shows the functional architecture of the energy management software.

6.1.2 Composition of the monitoring system

The microgrid monitors the DGs, ESs, loads, and control devices in the system. The monitoring system comprises photovoltaics (PV) monitoring,

Microgrid Technology and Engineering Application

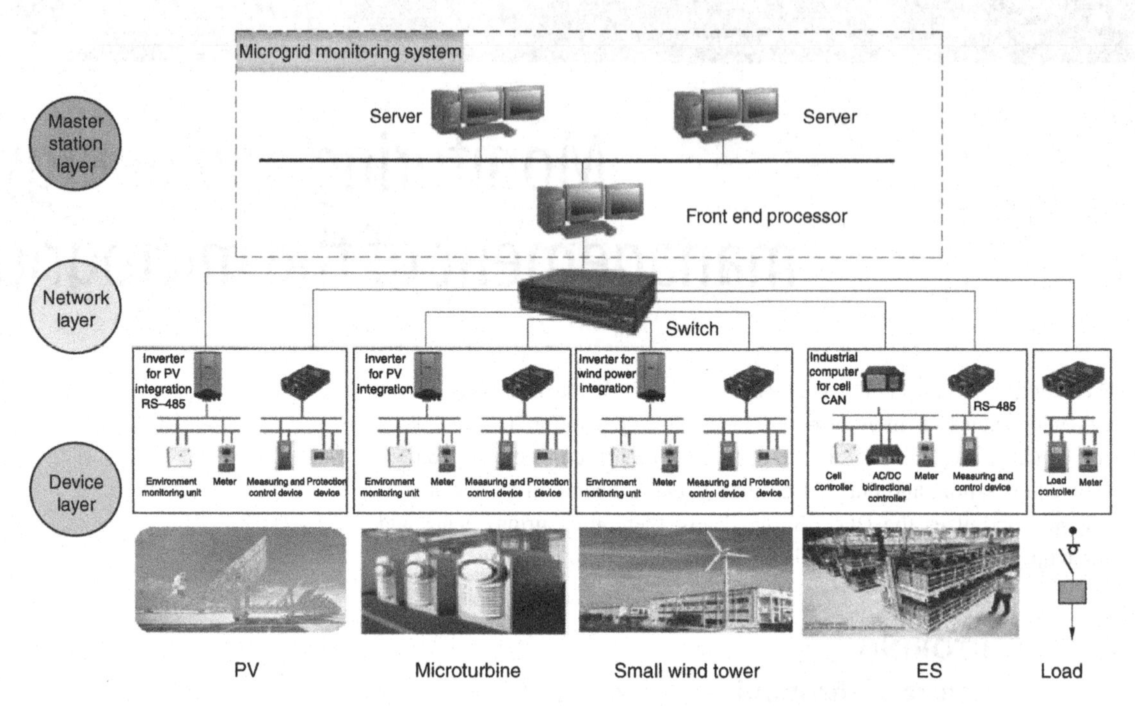

■ **FIGURE 6.1 Structure of the microgrid monitoring system.**

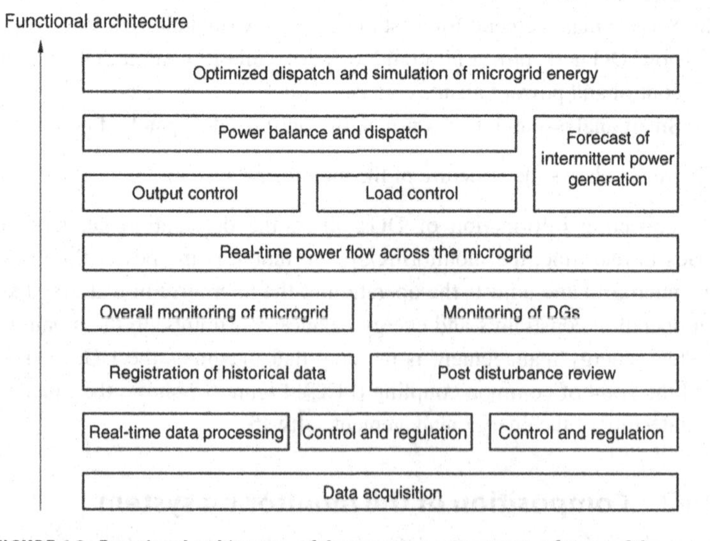

■ **FIGURE 6.2 Functional architecture of the energy management software of the microgrid monitoring system.**

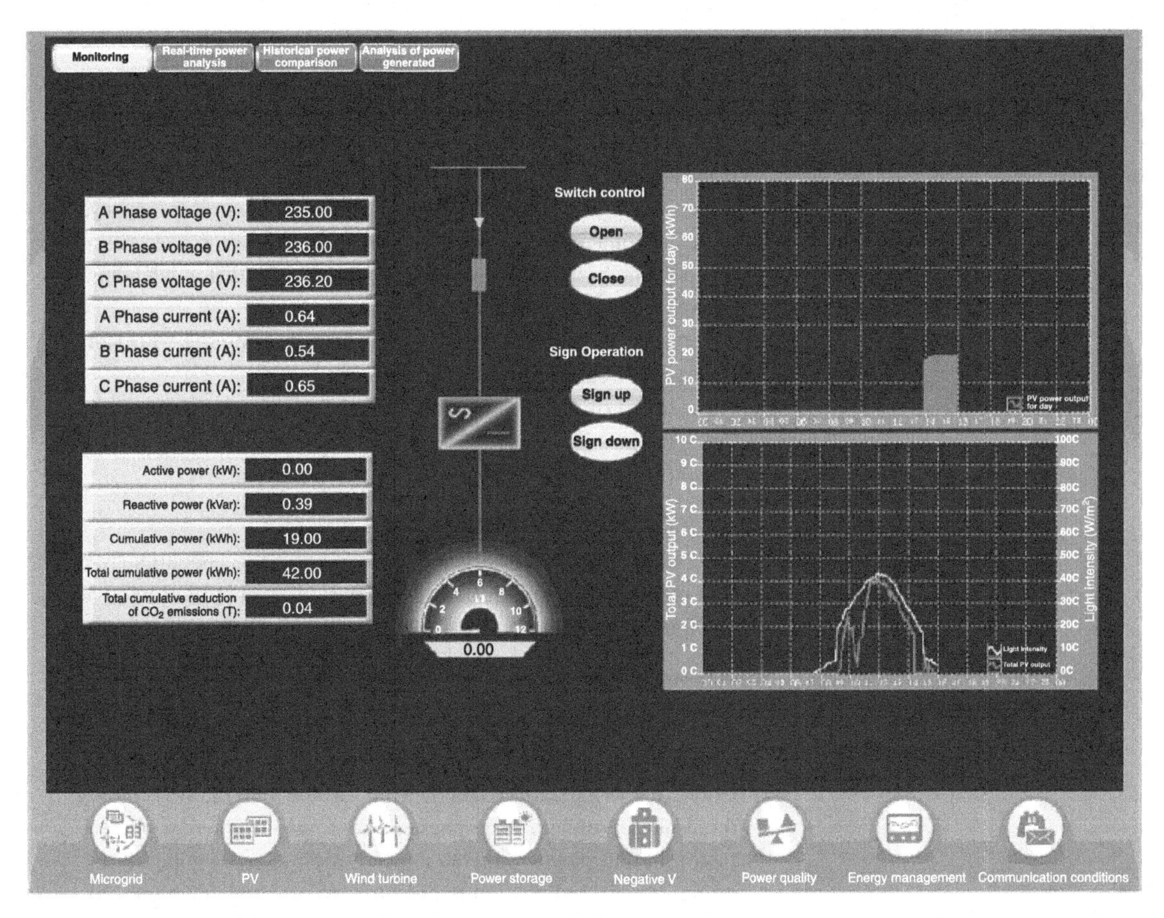

FIGURE 6.3 PV monitoring and statistics.

wind power monitoring, microturbine monitoring, monitoring of other power sources, ES monitoring, and load monitoring.

6.1.2.1 PV monitoring

The operation and alarm information of PV units is monitored in real time for comprehensive statistics, analysis, and control of PV electrification, as shown in Figure 6.3.

PV monitoring can achieve the following functions:

1. Real-time display of total power, daily total power output, aggregate power output, aggregate CO_2 emissions, and daily power output curve.
2. View of inverters' operation parameters, including DC voltage, direct current, DC power, AC voltage, alternating current, frequency, present power, power factor, daily power output, aggregate power output,

aggregate CO_2 emissions, inside temperature, and 24 h power output curve.

3. Monitoring of inverter operation, warning of equipment failure with audio and visual alarms, and identification of failure cause and occurrence time. Failure information includes excessively high grid voltage, low grid voltage, high grid frequency, low grid frequency, high DC voltage, low DC voltage, overload of inverter, overheating of inverter, short circuit of inverter, overheating of radiator, islanding of inverter, and communication failure.
4. Forecast of short-term and super-short-term power output, providing the basis for optimized dispatch of energy.
5. Adjustment of power output and control of start and stop of inverters.

6.1.2.2 Wind power monitoring

The operation and alarm information of wind turbine generators are monitored in real time for overall statistics, analysis, and control of wind power, as shown in Figure 6.4.

Wind power monitoring is mainly intended for the following:

1. Real-time display of total power, total daily power output, aggregate power output, and 24 h power output curve.
2. Collection of operation data of wind turbine generator, including three-phase voltage, three-phase current, grid frequency, power factor, output power, generator speed, rotor speed, temperature of generator windings, oil temperature in gearbox, ambient temperature, temperature of control board, wear and temperature of mechanical brake lining, cable twisting, nacelle vibration, anemometer, and wind vane.
3. Forecast of short-term and super-short-term power output, providing the basis for optimized dispatch of energy.
4. Adjustment of power output and control of start and stop of inverters.

6.1.2.3 Microturbine monitoring

The operation and alarm information of microturbines is monitored in real time for overall statistics, analysis, and control.

Microturbine monitoring is mainly intended for the following:

1. Monitoring of major operation parameters, including speed, gas inflow, gas pressure, exhaust pressure, exhaust temperature, knock intensity, and oxygen content;

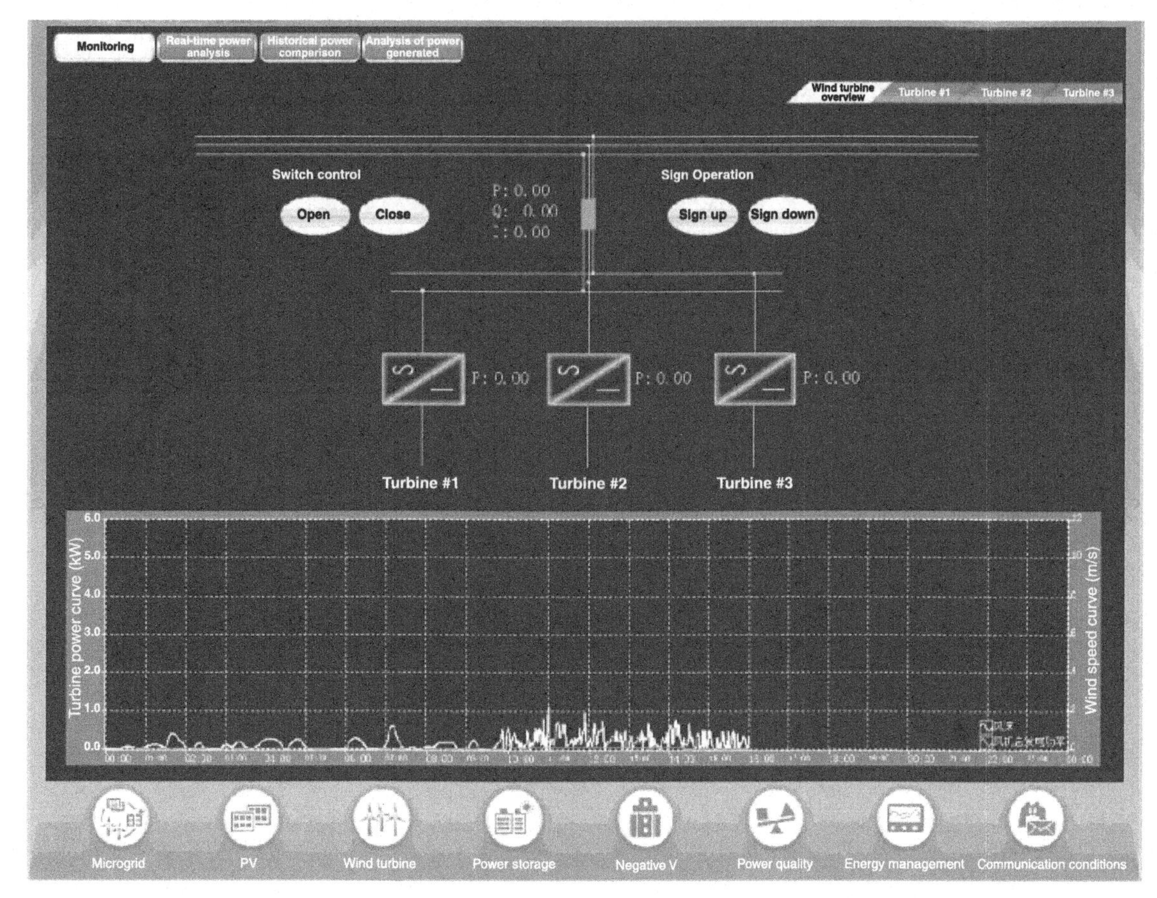

■ **FIGURE 6.4 Wind power monitoring and statistics.**

2. Monitoring of voltage, current, frequency, phase, and power factor before and after being connected to the grid;
3. Analysis, management, and adjustment of operation status.

6.1.2.4 Monitoring of other power sources

Similar to generation monitoring mentioned earlier, for other power sources, present output voltage of DG, working current, input power, grid-connected current, grid-connected power, grid voltage, current power output, aggregate power output, 24 h power output curve, and 24 h grid-connected power curve need to be monitored, for the purpose of ensuring secure and stable operation of the system.

6.1.2.5 ES monitoring

The operation and alarm information of batteries and power control system (PCS) is monitored in real time for overall statistics, analysis, and control of ES, as shown in Figure 6.5.

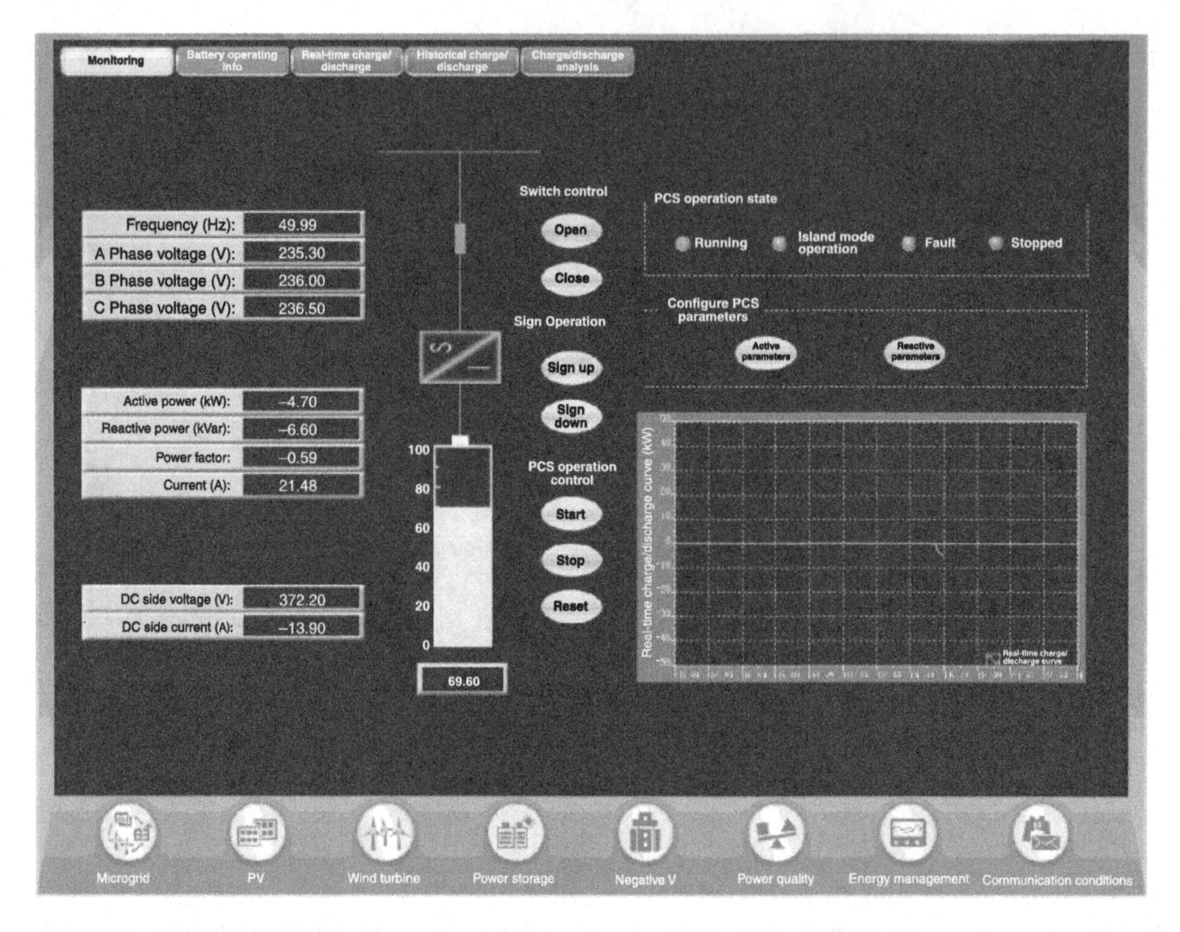

■ **FIGURE 6.5 ES monitoring.**

ES monitoring is mainly intended for the following:

1. Real-time display of energy that can be discharged or charged, maximum discharge power, present discharge power, discharge duration, total energy charged, and total energy discharged.
2. Remote communication of operation status, protection information, and alarm information of AC/DC bidirectional converter. Protection information includes under-voltage protection, over-voltage protection, open phase protection, under-frequency protection, over-frequency protection, over-current protection, component failure protection, battery failure protection, and overheating protection.
3. Remote measurement of battery voltage, battery charge, and discharge currents, AC voltage, and input and output power of the AC/DC bidirectional converter.

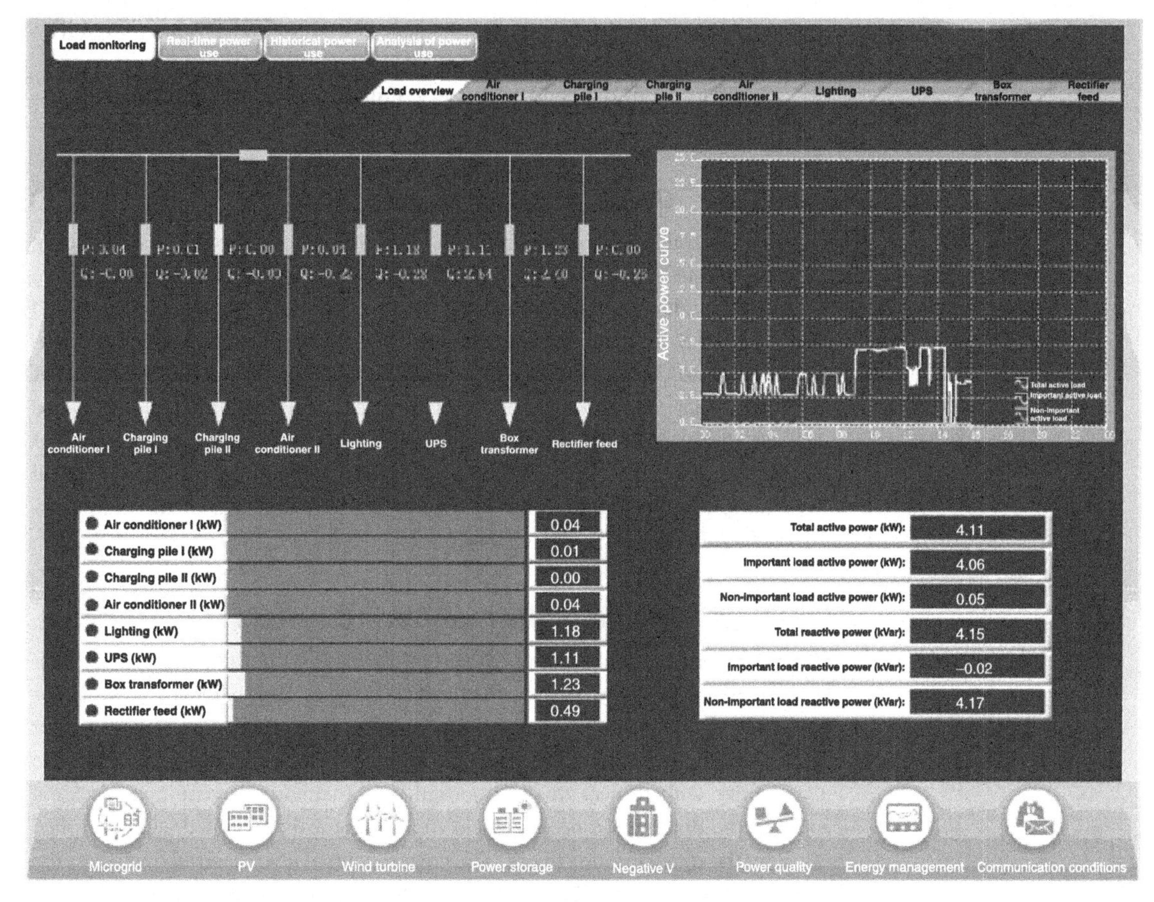

■ **FIGURE 6.6 Load monitoring.**

4. Remote regulation of battery charge and discharge duration and currents, and battery protective voltage, allowing for remote regulation of related parameters of the AC/DC bidirectional converter.
5. Remote control of battery charge and discharge of the AC/DC bidirectional converter.

6.1.2.6 *Load monitoring*

The operation and alarm information of loads is monitored for overall statistics, analysis, and control of loads, as shown in Figure 6.6.

Load monitoring is mainly intended for the following:

1. Monitoring of load voltage, current, active power, reactive power, and apparent power;
2. Recording of maximum load and occurrence time, maximum three-phase voltage and occurrence time, maximum three-phase power factor

and occurrence time, and statistics and monitoring of voltage eligibility rate and black-out occurrence time;

3. Warning of overload, and query of historical curve, reports, and events.

6.1.2.7 *Overall monitoring*

The overall operation information of the microgrid is monitored, including the frequency of the microgrid, voltage at the PCC, and power exchange with the distribution network; the total power output, state of charge (SOC) of the ES, total active loads, total reactive loads, total active power of sensible loads, total active power of controllable loads, and total active power of loads that can be completely shed are collected in real time; and the status of all circuit breakers in the microgrid, active power and reactive power of all branches, and alarm information of all equipment are monitored in real time, thus realizing real-time monitoring and statistics of the entire microgrid (see Figure 6.7).

6.1.3 Design of the monitoring system

The monitoring system of a microgrid should be designed to enable management and control of the distribution network dispatch layer, central control layer, and local control layer. The distribution network dispatch layer coordinates multiple microgrids (microgrid acts as a single controllable entity with respect to the macrogrid) to maintain security and economy of the distribution network, and the microgrid is regulated and controlled by the distribution network. The central control layer centrally manages the DG sources (including ES) and various types of loads, maximizes utilization and optimizes operation of the microgrid in grid-connected operation, and regulates the output of DG sources and load demands in islanded operation, thereby maintaining stability and security of the microgrid. The DG controller and load controller at the local control layer maintain transient power balance and under-frequency load shedding, ensuring transient security of the microgrid.

The monitoring system, a core link in the microgrid that integrates the DG, load, ES, and the PCC, generates control and regulation strategies based on fixed power balance algorithms to ensure stability of the microgrid in grid-connected or islanded operation and during transfer between various modes. Figure 6.8 exhibits the model of the energy management controller.

The coordination of microgrid local control and protection, microgrid central monitoring and management, and remote control of distribution and

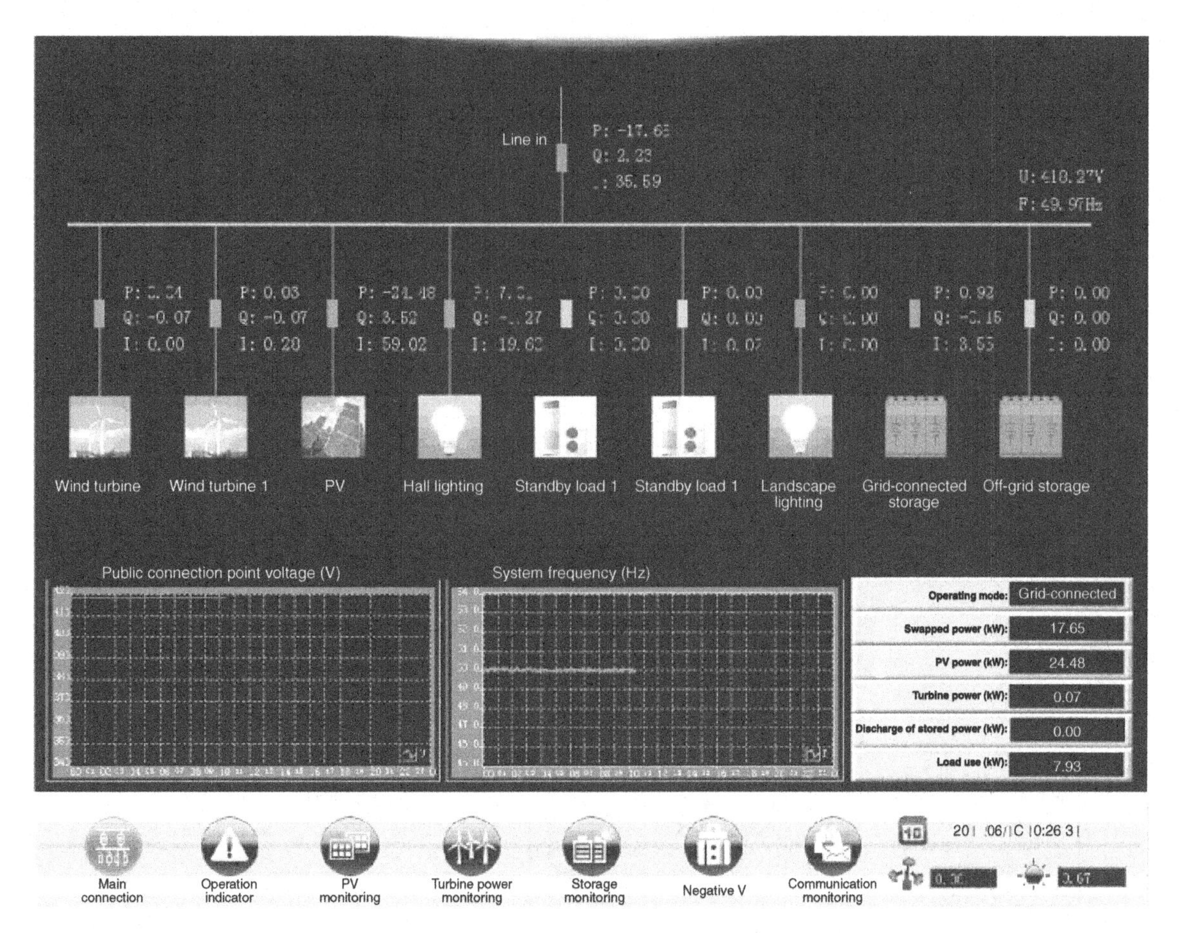

FIGURE 6.7 Overall microgrid monitoring and statistics.

dispatch can achieve power balance and control within the microgrid by control of the tie-line power flow. Figure 6.9 presents the coordinated control of an entire distribution system interconnected with the microgrid.

In addition to data acquisition, the following problems should be considered in designing the microgrid monitoring system to facilitate control, management, and operation of the microgrid:

1. *Protection of the microgrid*: Solutions should be recommended based on the rational settings of various protections of the microgrid and online check of rationality of protection settings, with a view to avoiding blackout due to unintentional operation of the protection under some circumstances.
2. *Integration of DG*: There are various types of DG sources in a microgrid that are distributed across the system and produce power intermittently. As such, the solution should provide ways for reasonable

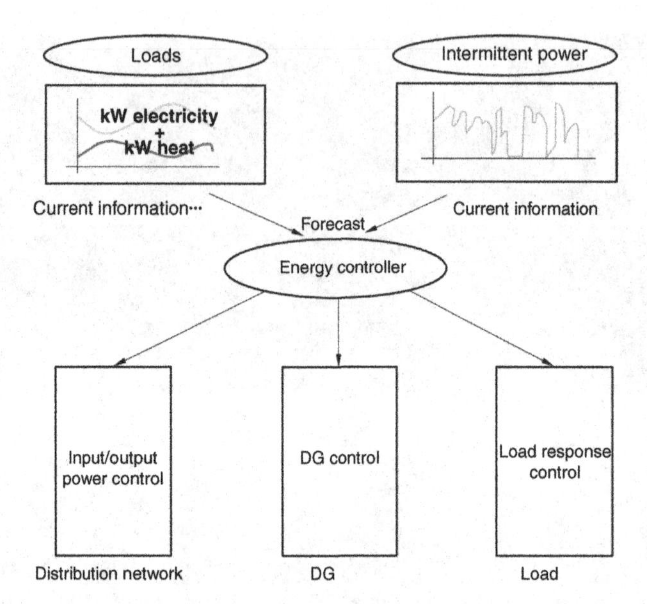

■ **FIGURE 6.8 Model of the energy management controller.**

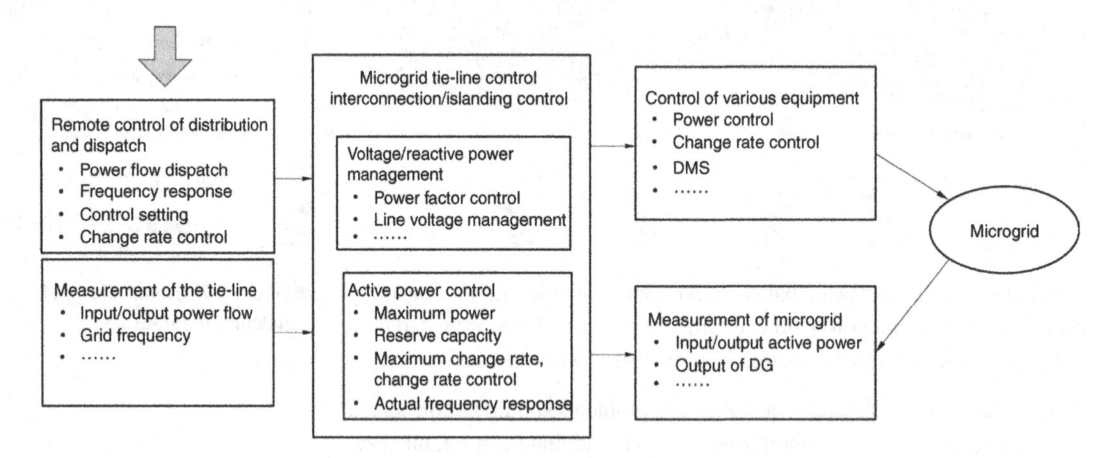

■ **FIGURE 6.9 Coordinated control of distribution system interconnected with microgrid.**

integration and coordination of DG sources to maintain stability of the microgrid both in grid-connected and islanded operations.

3. *Forecast of DG output*: The output of wind power and PV in a super-short-term is forecast based on the weather information released by the bureau of meteorology, historical meteorological information and historical records of power generation, thus enabling forecast and control of the microgrid.

4. *Microgrid voltage/reactive power balance and control*: The microgrid, as a single controllable power system, should meet the requirements placed by the distribution network on its power factor or absorption of reactive power to avoid long-distance transmission of reactive power, and should maintain a high voltage quality through voltage/reactive power balance and control when operating in parallel with the distribution network.
5. *Microgrid load control*: In islanded operation or when the distribution network has special requirements on loads or power output of the microgrid while the output of DG sources is fixed, the loads need to be shed, reconnected, or regulated based on their importance, so as to keep reliable power supply to important loads and maintain security of the microgrid.
6. *Microgrid power output control*: In islanded operation or when the distribution network has special requirements on loads or power output of the microgrid, the security and economy of the microgrid can be maintained by reasonably regulating the output of various DG sources, and in particular, by charging and discharging of batteries, and coordinating microturbines between cooling, heat, and power.
7. *Multilevel optimized dispatch of microgrid*: Load control and generation control are coordinated in various operation modes (grid-connected mode and islanded mode) and at various levels (DG, microgrid, and dispatch) to maintain the security and economy of the entire microgrid and provide support for optimized dispatch of the distribution network.
8. *Coordination between the microgrid and utility grid*: A microgrid can be deemed as a load or power source for the utility grid. Coordination between the two is conducive to reducing the loss of the distribution network and shifting loads. And when a serious fault occurs on the utility grid, the reasonable output of the microgrid will speed up the recovery of the grid.

6.2 ENERGY MANAGEMENT

Microgrid energy management is intended to maintain security and stability and improve energy efficiency of the microgrid by forecasting the DG, ES, and loads within the microgrid and optimizing control over the DG, ES, and loads based on their characteristics in grid-connected operation, islanded operation, and during transfer between different operation modes.

6.2.1 Forecast of DG

Forecast of DG, part of the microgrid energy management, is to forecast the short-term and super-short-term output of DG (wind power and PV) to provide a basis for optimized energy dispatch. It improves the use of DG sources, economic and social benefits, and increases reliability and economy of the microgrid.

The output of DG can be forecast by statistical methods or physical methods. The former is to find out the inherent law by collecting and analyzing historical data; and the latter to calculate using physical equations with meteorological data as the input.

Currently, continuous predication method, Kalman filtering method, random sequential method, artificial neural network method, fuzzy logic method, spatial correlation method, and support vector machine method are mainly used. Studies have been carried out on using these methods to forecast PV and wind power. In practice, high-precision ones should be selected from these methods considering their advantages and disadvantages.

With a high precision, the similar day and least squares support vector machine-based method can maintain the economy of the microgrid and meet the demands of control mode switching of the main power source for forecast of DG. This method involves two steps, selecting a similar day and forecasting the output of DG on the forecast day based on DG output on the similar day and weather data of the forecast day.

The similar day can be selected by level of correlation. The weather information released by the meteorological bureau includes the weather type, temperature, humidity, and wind strength. The days that have the same weather type (sunny, rainy, or cloudy) as the forecast day can be preliminarily selected as the similar day. The factors affecting PV power output are mainly irradiance and temperature, and those affecting wind power output mainly wind strength. The resemblance with the forecast day is then calculated from the most recent historical day, and the day most similar to the forecast day is used as the similar day. Finally, based on DG output on the similar day and the weather information of the forecast day, the DG output of the forecast day is calculated.

In the forecast of super-short-term DG output, after the data of DG output on the similar day is obtained, the DG output for the next hour can be forecast based on the weighted real-time weather data (irradiance, temperature, and wind strength) of the present hour.

6.2.2 Load forecast

Load forecast is to forecast future loads for analysis of demands, so that the operators can learn in time about the operation status of the system in the future. It is a major basis for the forecast of future operation of electrical power systems. Load forecast plays an important role in control, operation, and planning of the microgrid. Therefore, improving the forecast precision can contribute to a higher security and a better economy.

Load forecast methods used today are classified into traditional methods and modern methods based on their time of appearance. Traditional methods mainly include regression analysis method and sequential method, while modern methods mainly include expert system theory, neural network theory, wavelet analysis, gray system theory, fuzzy theory, and combinational method.

6.2.3 Frequency response characteristics of DG and loads

6.2.3.1 Response speed of DG

The DGs in a microgrid can be grouped as follows based on their frequency response capability and time:

1. *PV and wind*: Their output is affected by the weather, but not affected by the change in the system and hence, they can be deemed as constant-power sources.
2. *Microturbines and fuel cells*: Their response time ranges from 10 s to 30 s. In the case of a significant power deficiency in the microgrid system and the system has strict requirements on frequency. Instantly after islanding, microturbines and fuel cells cannot respond quickly enough to increase the output, and therefore, they are ignored in maintaining power balance at the instant of islanding.
3. The response time of ESs is generally 20 ms or even shorter. Therefore, it can be deemed that they can fill up the power deficiency with their maximum capacity in no time. The maximum capacity of ESs is roughly equal to the increment of power output that all DG sources can contribute to at the instant of islanding.

6.2.3.2 Frequency response characteristics of loads

The relationship between the active power of loads in an electrical power system and frequency of the system varies with the type of loads, as detailed next:

1. Loads whose active power does not vary with the frequency, such as lamp, electric furnace, and rectification loads;
2. Loads whose active power is in direct proportion to the frequency, such as ball mill, winch, compressor, and cutting machine;
3. Loads whose active power is in direct proportion to the square of frequency, such as the eddy current loss in the transformer core and feeder loss in the grid;

4. Loads whose active power is in direct proportion to the cube of frequency, such as ventilation fan and circulating pump with a small static head;
5. Loads whose active power is in direct proportion to the high degree of frequency, such as the feedwater pump with a large static head.

Without considering voltage fluctuation in the system, the relationship between system frequency f and the active power of load P_L can be expressed as

$$P_L = P_{LN}(a_0 + a_1 f_* + a_2 f_*^2 + \cdots + a_i f_*^i + \cdots + a_n f_*^n) \tag{6.1}$$

where $f_* = f/f_N$, N refers to the rated condition, * is the per unit value, P_{LN} is the active power of load at the rated frequency, and a_i is the scaling factor.

In the simplified system frequency response model, loads whose active power is in direct proportion to a high degree of frequency are not taken into account. Converting Eq. (6.1) to the differential equation of frequency, the frequency response factor of loads can be derived:

$$K_{L*} = a_{1*} = \frac{\Delta P_{L*}}{\Delta f_*} \tag{6.2}$$

Let ΔP be the power surplus and Δf be frequency increment, then

$$\begin{cases} \Delta P_{L*} = \dfrac{\Delta P}{P_{L\Sigma}} = \dfrac{\Delta P}{\Sigma P_{Li}} \\ \Delta f_* = \dfrac{\Delta f}{f_N} = \dfrac{f^{(1)} - f^{(0)}}{f^{(0)}} \end{cases} \tag{6.3}$$

where $f^{(0)}$ refers to the present frequency and $f^{(1)}$ the target frequency. In the case of power deficiency P_{qe} due to sudden change of the power output (e.g., tripping of generators) (if $P_{qe} < 0$, it indicates that more generators are switched in and power surplus occurs), part of loads will be shed to regulate the frequency, and then

$$K_{L*} = \frac{\Delta P_{L*}}{\Delta f_*} = \frac{\left(\dfrac{P_{qe} - P_{jh}}{P_{L\Sigma} - P_{jh}}\right)}{\left(\dfrac{f^{(1)} - f^{(0)}}{f^{(0)}}\right)} \tag{6.4}$$

where P_{jh} refers to the active power of loads to be shed. To reach the target frequency $f^{(1)}$ by load shedding, P_{jh} shall be

$$P_{jh} = P_{qe} - \frac{K_{L*}(f^{(1)} - f^{(0)})(P_{L\Sigma} - P_{qe})}{f^{(0)} - K_{L*}(f^{(1)} - f^{(0)})} \quad (6.5)$$

In the case of power surplus P_{yy} due to sudden change of loads (e.g., load shedding) (if $P_{yy} < 0$, it indicates that the number of loads rises and power deficiency occurs), some generators will be tripped and then

$$K_{L*} = \frac{\left(\dfrac{P_{yy} - P_{qj}}{P_{L\Sigma} - P_{yy}}\right)}{\left(\dfrac{f^{(1)} - f^{(0)}}{f^{(0)}}\right)} \quad (6.6)$$

According to Eq. (6.6), to reach the target frequency $f^{(1)}$, the active power of generators to be tripped P_{qj} is

$$P_{qj} = P_{yy} - \frac{K_{L*}(f^{(1)} - f^{(0)})}{f^{(0)}}(P_{L\Sigma} - P_{yy}) \quad (6.7)$$

6.2.4 Power balance

In grid-connected operation, generation and consumption in the microgrid are normally not limited, and only when needed, the macrogrid sends generation or consumption orders to the microgrid to control power exchange between the two. Specifically, in grid-connected operation, the macrogrid, based on economic analysis, sends power exchange setting values to the microgrid to maintain optimized operation. Based on the setting values, the energy management system of the microgrid will exercise control over the output of DG sources and charge and discharge of ESs, so that the microgrid operates on the specified power exchange rate in a secure and economic manner. In determining the output of various DG sources based on the power exchange setting values, the characteristics of the DG sources and control response characteristics should be considered for the energy management system.

6.2.4.1 Power balance in grid-connected operation

When the microgrid is grid-connected, the grid provides rigid voltage and frequency support, and normally exercises no special control over the microgrid.

In some cases, the grid specifies the amount of power exchange between the grid and microgrid, thus necessitating monitoring of power flow through the PCC.

When the actual power exchange deviates significantly from the setting value given by the grid, the microgrid control center (MGCC) needs to disconnect some loads or generators from the microgrid, or reconnect the loads or generators previously rejected to the microgrid, to minimize the deviation. The deviation of actual power exchange from the setting value is calculated as follows:

$$\Delta P^{(t)} = P_{\mathrm{PCC}}^{(t)} - P_{\mathrm{plan}}^{(t)} \tag{6.8}$$

where $P_{\mathrm{plan}}^{(t)}$ means the active power exchange setting value sent from the grid to the microgrid at the time of t, and $P_{\mathrm{PCC}}^{(t)}$ the active power flowing through the PCC at the time of t.

If $\Delta P^{(t)} > \varepsilon$, there is a power deficiency in the microgrid, and the MGCC needs to reconnect the generators previously tripped to the microgrid, or disconnect some less important loads from the microgrid; if $\Delta P^{(t)} < -\varepsilon$, there is a power surplus in the microgrid, the MGCC needs to reconnect the loads previously shed to the microgrid, or trip some DG sources that produce electricity at a higher cost.

6.2.4.2 *Power balance during transition from grid-connected mode to islanded mode*

At the instant of transition from grid-connected mode to islanded mode, the power flowing through the PCC is suddenly cut. If this power flows to the microgrid before the transition, a power deficiency of such an amount will occur in the microgrid after transiting to islanded mode; otherwise, a power surplus of such an amount will occur in the microgrid after the transition. The microgrid usually suffers a significant deficiency due to the sudden loss of power from the grid.

If, at the very beginning of islanded operation, emergency control measures are not taken, the microgrid will experience a dramatic frequency decline, causing protective outage of some DG sources, followed by a greater deficiency and further frequency decline, then protective tripping of other DG sources, and finally collapse of the microgrid. As such, to keep the microgrid in islanded operation for a long time, it is necessary to take control measures at the instant of the microgrid being separated from the grid to maintain power balance.

In the event of power deficiency at the very beginning of islanded operation, it is necessary to immediately shed all or some less important loads (or even some important loads) and increase the output of ES; in the event of power surplus, it is necessary to immediately reduce the output of ES or even trip some of the DG sources. This will restore the microgrid to power balance quickly.

The instant deficiency (or surplus, as the case may be) in the microgrid is equal to the power flowing through the PCC before the separation.

$$P_{qe} = P_{PCC} \tag{6.9}$$

P_{PCC} is expressed as a positive number if the power flows from the grid to the microgrid and vice versa. A P_{qe} greater than 0 indicates power deficiency in the microgrid at the instant of separation; and a P_{qe} smaller than 0 indicates power surplus.

As the ESs are intended to provide uninterrupted power supply to important loads for a certain period in islanded mode, such principles for power balance at the instant of separation apply that less important loads are shed first under the assumption that the output of all ESs is 0, then the output of the ESs is adjusted, and finally some important loads are shed if necessary.

6.2.4.3 **Power balance in islanded operation**

The microgrid is capable of operating in both grid-connected mode and islanded mode. When the microgrid disconnects from the grid after a fault, by adjusting the output of DG sources, output of ESs, and loads to achieve power balance and control, the microgrid can maintain stable operation. This ensures uninterrupted supply to important loads with DGs, thereby contributing to a higher efficiency of DG sources and supply reliability.

During islanded operation, the output of DG sources in the microgrid may vary with the environment (such as the irradiance, wind strength, and weather condition), leading to significant fluctuations of voltage and frequency. Therefore, it is necessary to monitor the voltage and frequency of the microgrid in real time, so that measures can be taken in time to deal with sudden change of sources and load that may impair the security and stability of the microgrid.

Supposing that the power deficiency at a moment during islanded operation is P_{qe}, then $\Delta P_L * = P_{qe} / P_{L\Sigma}$. It can be inferred from Eq. (6.2) that

$$P_{qe} = \frac{f^{(0)} - f^{(1)}}{f^{(0)}} \times K_{L*} P_{L\Sigma} \tag{6.10}$$

If, in islanded operation, the frequency at a moment $f^{(1)}$ is lower than f_{min}, there occurs a power deficiency on the microgrid, requiring the MGCC to reconnect the generators previously tripped, or shed part of the less important loads. While if $f^{(1)}$ is higher than f_{max}, there appears a remarkable power surplus on the microgrid, requiring the MGCC to reconnect loads previously shed or trip some DG sources.

1. Load-shedding control in the case of power deficiency
   In the case of power deficiency ($P_{qe} > 0$), the following control strategies apply:
   a. Calculate the current active output $P_{S\Sigma}$ and maximum active output P_{SM} of ESs.

$$\left.\begin{aligned} P_{S\Sigma} &= \sum P_{Si} \\ P_{SM} &= \sum P_{Smax-i} \end{aligned}\right\} \tag{6.11}$$

   where P_{Si} is the active output of the ES i, which is positive during discharge and negative during charge.
   b. If $P_{qe} + P_{S\Sigma} \le 0$, it indicates the ES is being charged, and if the charging power is greater than the power deficiency, reduce the charging power until $P'_{S\Sigma} = P_{S\Sigma} + P_{qe}$, and stop the control. Otherwise, set the active output of the ES to 0 and recalculate the power deficiency P'_{qe}.

$$\left.\begin{aligned} P'_{qe} &= P_{qe} + P_{S\Sigma} \\ P_{S\Sigma} &= 0 \end{aligned}\right\} \tag{6.12}$$

   According to Eq. (6.5), the allowable forward and reverse deviations of power deficiency can be calculated based on the maximum frequency f_{max} and minimum frequency f_{min}:

$$\left.\begin{aligned} P_{qe+} &= \frac{K_{L*}(f_{max} - f^{(0)})(P_{L\Sigma} - P_{qe})}{f^{(0)} - K_{L*}(f_{max} - f^{(0)})} \\ P_{qe-} &= \frac{K_{L*}(f^{(0)} - f_{min})(P_{L\Sigma} - P_{qe})}{f^{(0)} + K_{L*}(f^{(0)} - f_{min})} \end{aligned}\right\} \tag{6.13}$$

   c. Determine the amount of less important loads to be shed

$$\left.\begin{aligned} P^{(1)}_{jh-min} &= P_{qe} - P_{qe-} \\ P^{(1)}_{jh-max} &= P_{qe} + P_{qe+} \end{aligned}\right\} \tag{6.14}$$

   d. Shed less important loads. Shed loads in an ascending order of importance. For loads of the same importance, shed them in a descending order of power. If P_{Li} (power of a load) > $P^{(1)}_{jh-max}$, do not shed this load and proceed to check the next one; if $P_{Li} < P^{(1)}_{jh-min}$, shed this load and proceed to check the next one. If $P^{(1)}_{jh-min} \le P_{Li} \le P^{(1)}_{jh-max}$, shed this load and stop checking other loads. After shedding the load i, recalculate the power deficiency based on Eq. (6.15), and the amount of less important loads needing to be shed based on Eq. (6.14), and then proceed to check the next load.

$$P'_{qe} = P_{qe} - P_{Lqc-i} \tag{6.15}$$

where P_{Lqc-i} means the active power of loads that are shed.

e. After shedding less important loads as appropriate, if $-P_{SM} \le P_{qe} \le P_{SM}$, adjust the output of ESs to provide the remaining power deficiency until $P_{S\Sigma} = P_{qe}$, and then stop the control. Otherwise, calculate the amount of important loads needing to be shed, that is

$$\left.\begin{aligned} P^{(2)}_{jh-min} &= P_{qe} - P_{SM} \\ P^{(2)}_{jh-max} &= P_{qe} + P_{SM} \end{aligned}\right\} \tag{6.16}$$

f. Shed important loads in a descending order of power. If P_{Li} (power of a load) $> P^{(2)}_{jh-max}$, do not shed the load and proceed to check the next one; if $P_{Li} < P^{(2)}_{jh-min}$, shed the load and proceed to check the next one; if $P^{(2)}_{jh-min} \le P_{Li} \le P^{(2)}_{jh-max}$, shed the load and stop checking other loads. After shedding the load i, recalculate the power deficiency based on Eq. (6.15) and the amount of important loads needing to be shed based on Eq. (6.16), and then proceed to check the next load.

g. Adjust the output of ESs to provide the remaining power deficiency after appropriate load shedding until $P_{S\Sigma} = P_{qe}$.

2. Generator tripping control in the case of power surplus

In the case of power surplus ($P_{yy} > 0$), it is necessary to trip some generators, and the control strategies are similar to those in the case of power deficiency:

a. Calculate the current and maximum active output of the ESs based on Eq. (6.11).

b. If $-P_{SM} \le P_{yy} - P_{S\Sigma} \le P_{SM}$, adjust the output of the ESs to absorb the power surplus after a proper number of generators are tripped until $P'_{S\Sigma} = P_{yy} - P_{S\Sigma}$, and then stop the control. Otherwise, proceed to the next step.

c. Calculate the allowable forward and reverse deviations of power surplus based on the allowable upper and lower limit of frequency:

$$\left.\begin{aligned} P_{yy+} &= \frac{K_{L*}(f^{(0)} - f_{min})}{f^{(0)}}(P_{L0} - P_{yy}) \\ P_{yy-} &= \frac{K_{L*}(f_{max} - f^{(0)})}{f^{(0)}}(P_{L0} - P_{yy}) \end{aligned}\right\} \tag{6.17}$$

d. If the ES is being discharged ($P_{S\Sigma} > 0$), set the discharge power to 0 and recalculate the power surplus:

$$\begin{cases} P_{yy} = P_{yy} - P_{S\Sigma} \\ P_{S\Sigma} = 0 \end{cases} \tag{6.18}$$

e. Calculate the amount of DG sources needing to be tripped:

$$\left.\begin{aligned} P_{qj-min} &= P_{yy} - P_{SM} - P_{S\Sigma} - P_{yy-} \\ P_{qj-max} &= P_{yy} + P_{SM} - P_{S\Sigma} + P_{yy+} \end{aligned}\right\} \tag{6.19}$$

f. Trip the generators in a descending order of power. If P_{Gi} (power of a source) > $P_{qj\text{-}max}$, do not trip the source and proceed to check the next one; if $P_{Gi} < P_{qj\text{-}min}$, trip the source and proceed to check the next one; if $P_{qj\text{-}min} \le P_{Gi} \le P_{qj\text{-}max}$, trip the source and stop checking other sources. After tripping the source *i*, recalculate the power surplus based on Eq. (6.20) and the amount of sources needing to be tripped based on Eq. (6.19), and proceed to check the next source.

$$P'_{yy} = P_{yy} - P_{Gqc-i} \tag{6.20}$$

where P_{Gqc-i} means the active power of DG that is tripped.

g. Adjust the output of ESs to absorb the remaining power surplus after a proper number of generators are tripped until $P_{S\Sigma} = -P_{yy}$.

6.2.3.4 Power balance

After the microgrid is reconnected to the grid, the DGs switch to *P*/*Q* control, and their power output relies on the dispatch plan of the distribution network. The MGCC needs to gradually put the loads or generators that were automatically disconnected from the microgrid into operation to maintain security and stability of the microgrid.

6.3 OPTIMIZED CONTROL

Renewable sources, such as wind power and PV power, are connected to the grid via inverters. Normally, the inverters are controlled to follow the frequency and voltage of the system, and keep the maximum output of DG sources rather than adjust their own output when the system frequency or voltage is excessively high or low. To maintain stability, there must be a master power source that can automatically change its output following the change of power frequency and voltage. Either a rotating generator, such as a diesel generator or pumped storage unit, or a large-capacity ES can serve as a master power source. Renewable sources shall be used as practical as possible. When renewable power is insufficient, the master power source can be used.

The economic operation control of the microgrid aims to maximize energy efficiency, minimize operation costs, and ensure the best economy of the microgrid by making best use of renewable energy while ensuring grid stability. Various optimization measures have been developed in view of the characteristics of various sources.

6.3.1 Optimized control of PV power

Optimized control of PV power is considered from the characteristics of PV sources and characteristics of power generation equipment.

1. *Control of maximum output*: PV power is a type of renewable energy that is relatively stable and has the highest priority among all renewable sources. PV units usually work at the maximum power output, except when output exceeds the demand, and the ESs have been fully charged.
2. *Control of inverter group*: PV inverters have the best efficiency and power quality when working at 30–70% of their rating. The efficiency of a PV inverter is dependent on the input power. When the input power is much less than the rated power (e.g., less than 20% of the rating), the efficiency will drop significantly. Furthermore, the total harmonic distortion (THD) in the output current of a PV inverter decreases with the increase of the input power. When inverters are lightly loaded, the THD will increase significantly. Specifically, the THD will exceed 5% when the input power is less than 20% of the rated power, and may even exceed 20% in a less-than 10% case. However, the efficiency will also decline when the input power is more than 80% of the rated power. As such, the PV inverter group control is used in the microgrid to improve the overall efficiency of the PV system.

   The PV inverter group control divides PV arrays into groups, and distributes the current on the DC side of the PV system to more inverters or a centralized inverter through a transfer switch. The control strategy is as follows: in the morning, depending on sunlight availability, one inverter is started first, and when the inverter is almost fully loaded, another one is started. Then, in this way, other inverters are started one by one. In the evening, inverters are shut off one by one according to the output of solar panels. This kind of control requires prediction of sunlight change and rain and detection of the input power throughout the day. When the input power is too low, the transfer switch on the DC side is controlled to collect all DC current to one inverter, so that the efficiency of inverters will not decrease significantly due to the decrease of the input power.

6.3.2 Optimized control of wind power

Owing to the uncertainty of wind resources and operation characteristics of wind turbine generators, the output of wind turbine generators fluctuates remarkably, which often causes voltage deviation, fluctuation, and flicker. Thus, the primary concern for wind power control of a microgrid is to eliminate the influence on the stability due to the output fluctuation.

The following solutions are proposed to solve this problem:

1. The MGCC obtains the global measurement data of the system. When significant frequency or voltage fluctuation is detected, the controller determines and issues the control scheme immediately to compensate for the fluctuation to maintain constant power output.
2. Where the power of ESs is limited (the SOC is too small or too large, for instance), the energy management system of the microgrid may send power output orders to wind power inverters to temporarily control the output of wind power. As long as the designated output power varies slowly, the wind turbine generators can maintain a constant output. This limits the maximum output of wind turbine generators. After the generators resume a constant output, their output is increased to the maximum by control means to fully utilize wind resources. When renewable output exceeds the demand, for stability concerns, wind turbine generators are taken out of service first to maintain maximum PV output.

6.3.3 Optimized control of various types of ESs

ESs play an important role in keeping the stability of a microgrid. Usually, depending on the actual demand, diversified ESs are provided. An optimized dispatch system has to develop different control strategies for different types of ESs according to their respective characteristics.

1. *Battery ES system (energy type)*: This type of ES is featured with a small loss, a long storage period, but a low response speed and a short life cycle, and is used only for storage of a large amount of renewable energy and as backup power sources for loads.
2. *Flywheel, super-capacitor, and SMES (power type)*: This type of ES is featured with a high response speed and output power, but a large self-loss in the storage process, and is unsuitable for long-time storage.

Therefore, they are mostly used in such circumstances as emergency power deficiency, mode transfer, and system disturbance.

Optimized control of multiple types of ESs can maintain smooth power output and stable voltage and provide backup in emergency. Whenever using ESs, their SOC should be watched all the time. If the SOC is excessively low or high, no power output orders should be given to them to prevent over-charge or over-discharge.

6.3.4 Optimized dispatch strategies

6.3.4.1 *Power exchange management in grid-connected mode*

In grid-connected operation, generation and consumption in the microgrid are normally not limited, except that the macrogrid, when necessary, sends specific generation or consumption orders to the microgrid through power exchange control. That is, in grid-connected operation, the macrogrid sends power exchange settings to the microgrid for best economy according to analysis results. According to the setting, the energy management system of the microgrid controls the output of DG sources and charge or discharge of ESs to maintain power exchange as instructed while ensuring economic and secure operation of the microgrid. In determining the output of each DG source, the energy management system should take the characteristics and control response characteristics of various DG sources into account.

6.3.4.2 *Energy balance control in islanded operation*

When the microgrid switches to islanded operation following a fault on the macrogrid, it should be able to maintain stability by energy balance control. In islanded mode, energy balance control, by adjusting the output of DG sources, energy release of ESs, and power consumption of loads, can maintain stability of the microgrid and ensure continuous power supply to important loads while fully utilizing renewable power, thereby improving efficiency of DG sources and supply reliability.

Chapter 7

Communication of the microgrid

The microgrid is operated, controlled, and managed in a different way from conventional grids mainly in that it relies more on information collection and transmission. The response characteristics of microgrid equipment pose higher requirements on timeliness and reliability of the communication system. The communication system is vital for managing and controlling the microgrid.

A microgrid effectively combines flow of power, service, and information. Real-time information collection, timely and reliable data transmission, and efficient processing and intelligent analysis of multilevel data can meet requirements for timeliness, accuracy, and comprehensiveness of data.

7.1 SPECIAL REQUIREMENTS ON COMMUNICATION

Different from common communication systems, the communication system of a microgrid is expected to meet a wide range of requirements for the bandwidth, timeliness, reliability, and security. The requirements on the communication system are special in the following four aspects.

7.1.1 High integration

The high integration of the communication system is expressed in technologies and services. The communication system integrates computer network technology, control technology, sensing, and metering technology, and can connect the microgrid with various power communication networks (switched telephone network, power data network, relay protection network, teleconference network, enterprise intranet, and security system), thus realizing seamless communication from generation to consumption, allowing for access of various types of generators and ESs, simplifying the integration process, and making possible the "plug and play" of the microgrid service and application and automatic control of the microgrid.

Microgrid Technology and Engineering Application

7.1.2 High reliability

Unlike a robust macrogrid, a microgrid is relatively weak. This requires the microgrid to have fast recovery and automatic control, which therefore calls for a highly reliable communication system. When a fault occurs on the microgrid, the communication system should quickly isolate the fault, switch loads to reliable power sources, and collect key data of the faulty section to reduce the downtime after a severe fault.

7.1.3 Commonly recognized standards

To achieve bidirectional, real-time, and efficient communication, the communication system must be established based on open and common standards, which may support high-speed and accurate communication between sensors, advanced electronics, and applications.

7.1.4 Higher cost-effectiveness

The communication system facilitates the operation of a microgrid in such a way that through predication, it prevents the occurrence of events adversely affecting the reliability of the grid to avoid cost rise due to poor power quality, and the microgrid-based automatic communication surveillance function also significantly reduces costs for personnel monitoring and equipment maintenance.

7.2 DESIGN PRINCIPLES OF THE COMMUNICATION SYSTEM

A microgrid shall be provided with a high-speed, bidirectional, broadband, and automatic communication system to allow for flexible access of multiple services and "plug and play" communication. As such, the following principles should be observed in designing the communication system of a microgrid.

7.2.1 Unified planning and design

The communication system is not only the basis for control and operation of the microgrid, but also a prerequisite to make its commercialization possible. Therefore, the communication system should be open, scalable, secure, open, and coordinated with the services and commonly recognized communication standards.

7.2.2 Secure, reliable, and open

Given the great variety and frequent interaction of users of a microgrid, the design of the communication system shall not only meet the requirement for openness, but also ensure the security of major equipment and users' privacy.

7.2.3 Scalability

The increasing penetration of DGs and loads and amount of data place a higher requirement on the bandwidth and reliability of the transmission network. This should be fully considered in designing the communication system by providing redundancy for network extension and maintenance and upgrade.

7.3 COMMUNICATION SYSTEM OF THE MICROGRID

7.3.1 Communication technologies

At present, quite a lot of communication technologies are available and they can be roughly classified into wired communication and wireless communication. Wired communication includes optical fiber communication and power line communication (PLC), while wireless communication includes spread spectrum communication, WLAN (IEEE 802.11), WWAN (IEEE 802.20), GPRS/CDMA, 3G/4G, satellite communication, microwave, shortwave/ultrashortwave, and space optical communication. The following briefly introduces some of these technologies.

7.3.1.1 Optical fiber communication

Optical fiber communication, including traditional communication, Ethernet communication, and passive optical network communication, is advantageous in high communication capacity, low loss, long transmission distance, high resistance to electromagnetic interference, and high transmission quality and speed.

Ethernet passive optical network (EPON) is a point-to-multipoint optical fiber transmission and access technology adopting the broadcast mode in the downlink and time division multiple access (TDMA) mode in the uplink. It can be flexibly configured in a tree, star, or bus topology, and at the branching points, except for an optical splitter, no node equipment is required, contributing to saving of optical fiber resources, sharing of bandwidth, saving of investment in computer room, high equipment security, easy networking, and low networking cost. EPON is best suited for point-to-multipoint communication, in which a passive optical splitter alone can achieve distribution of optical power.

EPON has such advantages as simple equipment, low installation and maintenance cost, low investment, flexible networking in tree topology, star topology, bus topology, hybrid topology, redundant topology, and ease of configuration. It can be used indoors or outdoors.

7.3.1.2 PLC

PLC is a promising broadband access technology. In this transmission mode, multimedia service signals, including high-speed data, voice, and video, are transmitted over low-voltage power lines.

PLC has such advantages as the power line is under the full jurisdiction of the power sector and easily managed; it can connect to any measurement and control point; signals can be transmitted through power lines, obviating the need to erect special lines; and no permission from the Federal Communications Commission is necessary.

It also has some disadvantages, such as low transmission speed; sensitivity to disturbance, nonlinear distortion, and cross-modulation between channels; large size, and high price of capacitors and inductors used in the PLC system.

7.3.1.3 PTN

Packet transport network (PTN) is a new-generation transport network oriented to packet data services. It is a layer configured between the service layer with Ethernet as the external manifestation and optical fiber transmission layer such as wavelength division multiplexing. It is intended to meet the sudden increase and statistical multiplexing transport requirements of IP services, and with packet services as the core, supports transport of multiple services. This technology not only contributes to a lower overall cost, but also inherits the advantages of synchronous digital hierarchy (SDH). Its functions are totally customized based on IP service transport requirements. It has the following features:

1. A connection-based technology, meeting requirements for carrier-class services. With high quality of service, identification and differentiated management of users, and prioritizing of services and bandwidth management, this kind of flexible connection management provides more management modes, more user access options, and higher cost efficiency of statistical multiplexing than traditional circuit-based connection.
2. Higher operation administration and maintenance (OAM) capability. All PTN technologies emphasize end-to-end OAM capability. Specifically, in addition to the connection management and loopback means required for traditional packet transport devices, PTN technologies also enhance performance management to meet requirements for carrier-class services, such as packet loss rate and time delay measurement. Another significant feature is tandem connection monitoring.

3. Rapid network protection. PTN provides linear protection switching and ring network protection, and in particular, point-to-point channel protection switching within 50 ms.

The PTN technology that integrates the native time division multiplex (native TDM) and packet service is the best solution to data transmission. It not only meets the high requirements for timeliness of relay protection and telecontrol, but also supports the evolution of the operation management platform to packet services. All in all, PTN can ensure reliable communication service for the microgrid.

7.3.2 Structure

The communication system of a microgrid is used throughout power generation, transmission, and distribution, and can be divided into two parts based on the scope of application: (1) the monitoring communication network for the microgrid and (2) the communication network between the control center of the microgrid and that of the distribution network.

7.3.2.1 Microgrid monitoring communication network

Figure 7.1 shows the architecture of the network. With advanced communication technologies, the network works for power dispatch, on-line real-time monitoring of power equipment, management of site operation videos, energy information collection, and anti-theft of outdoor facilities. Communication is mainly through electrical optical fiber network, PTN, and WLAN.

7.3.2.2 Communication network between the control center of microgrid and that of the distribution network

The network is generally set up by referring to the architecture of the communication network of a smart distribution network, in which the microgrid is regarded as an active, controlled client.

7.3.3 Design

In designing the communication system of a microgrid, the communication technologies should be reasonably selected and mixed by fully considering such factors as their respective advantages and applications, costs, and ambient environment, so that the advantages of various technologies can be fully utilized. The communication technologies are selected mainly based on the type of data to be transmitted, geographical distribution of communication nodes, and scale of the microgrid.

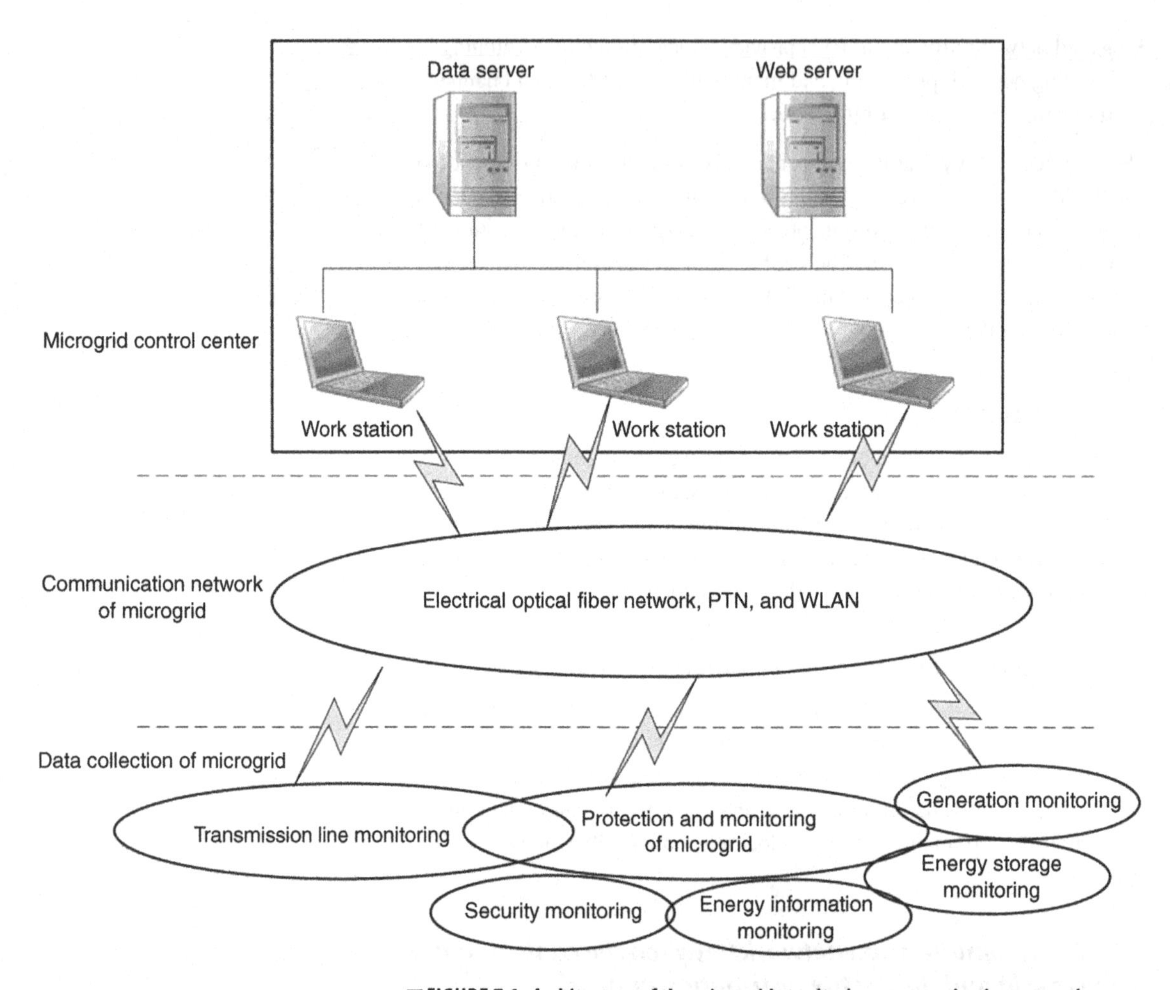

FIGURE 7.1 Architecture of the microgrid monitoring communication network.

7.3.3.1 Functional model of the communication system

The communication system of a microgrid is intended for bidirectional, timely and reliable transmission of control, monitoring, and user data. It is a comprehensive platform integrating communication, information, and control technologies. Figure 7.2 shows the communication flowchart of a microgrid.

The control information of generation, transmission, transformation, distribution, and dispatch of the microgrid is transmitted through the dispatch data network, while the control information of protection, safety, and stability, which has a stringent requirement for time delay, is transmitted through special lines.

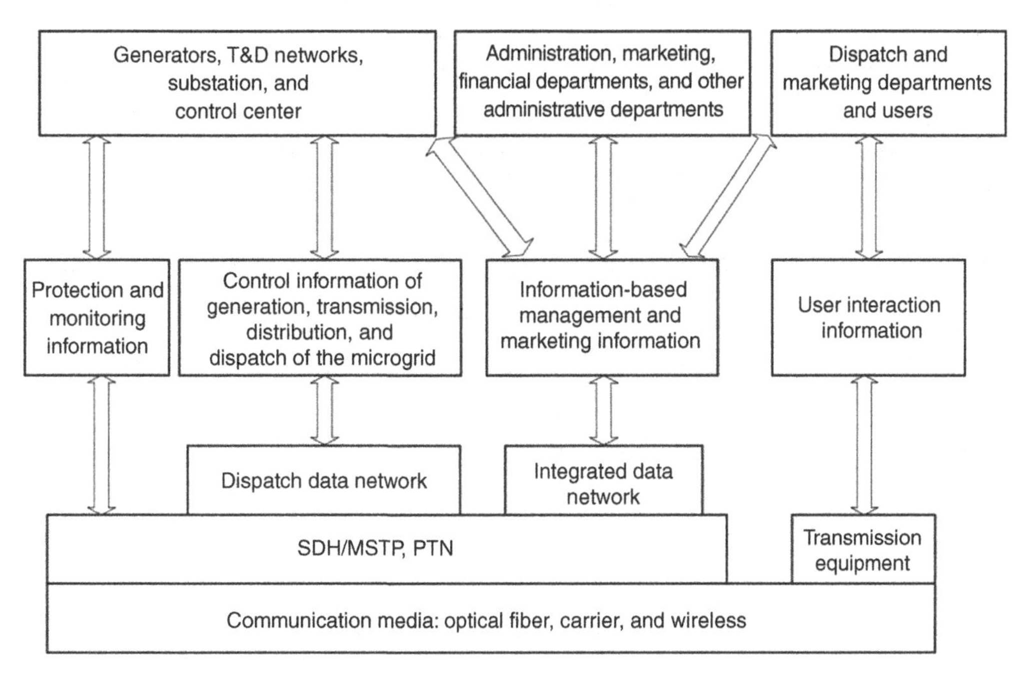

FIGURE 7.2 Communication flowchart of microgrid. SDH, synchronous digital hierarchy; MSTP, multiservice transport platform.

User information is communicated with technologies that suit the user network access characteristics of the power system and meet interaction requirements.

Information services of the management, operation, maintenance, and sales departments and other administrative departments are carried over the integrated data network, and with the development of the microgrid, voices and other private line services will also be carried over the integrated data network.

7.3.3.2 ***Typical solution to communication of an insular microgrid***

Figure 7.3 shows the structure of a typical insular microgrid. The communication system can be divided into three layers by function: (1) local control layer, (2) central control layer, and (3) distribution network dispatch layer. The local control layer, consisting of various data collection devices directly connected with the measured objects, transfers the measured data to and receives orders from the central control layer; the central control layer receives measured node information from the local control layer, gives control

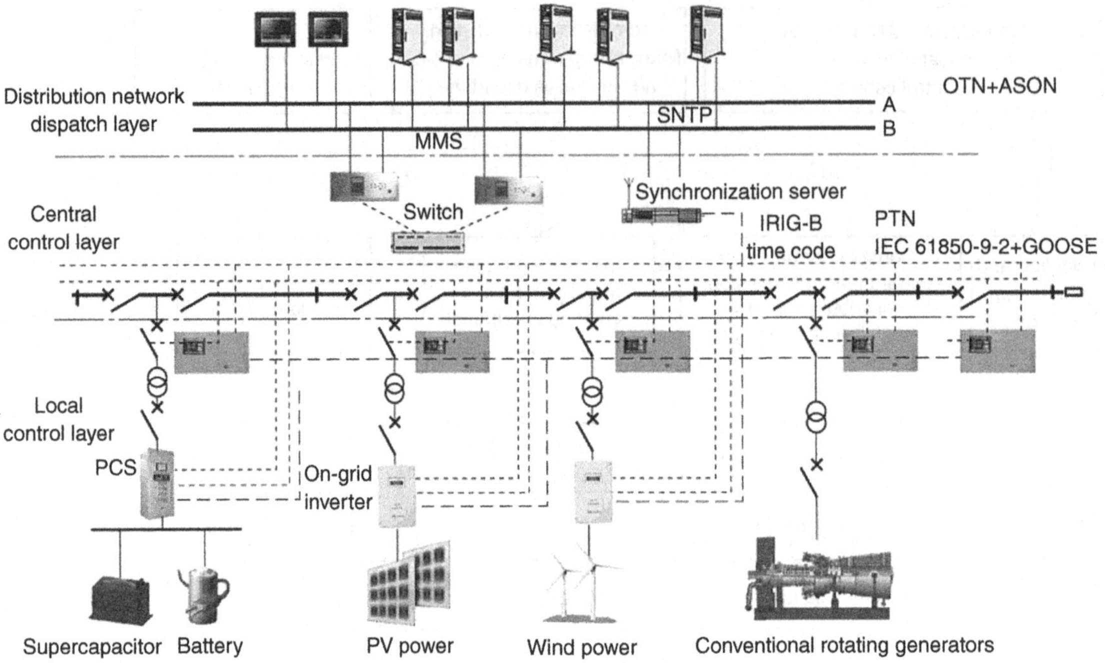

■ FIGURE 7.3 Structure of an insular microgrid. MMS, manufacturing message specification; OTN, optical transport network; SNTP, simple network time protocol; ASON, automatic switched optical network.

orders to the local control layer, and receives instructions from the distribution network dispatch layer and then transfers the information to the local control layer; the distribution network dispatch layer receives data from the central control layer, analyzes the data, and makes decisions on control of the microgrid.

1. *Communication design of the local control layer*: The microgrid is a system combining loads and power sources. The local control layer should have such functions as collection of microgrid data, local protection and control of equipment, high-precision and quick collection of measured data of feeders, protection of feeders against failure, monitoring of inverters for grid connection of DGs, maximum power point tracking, grid dispatch mode, control of automatic and seamless transfer between *P*/*Q* mode and *U*/*f* mode of power converter system, and hybrid access of ESs (lithium iron phosphate battery, lead acid battery) and supercapacitor. These functions necessitate support of a strong communication system. Various parts of the microgrid should have a unique communication architecture that corresponds to their respective application and security demands.

   For a small microgrid with a few nodes and a simple network structure, the communication protocols of the collection devices (e.g., RS-485, real-time industrial Ethernet) may be directly used; for a large microgrid with many nodes and a complicated network structure, optical fiber transmission may be adopted.

2. *Communication design of the central control layer*: The central control layer mainly serves dynamic power disturbance control and central fault protection, including central differentiated protection of the microgrid, control of transition between grid-connected mode and islanded mode, quick dynamic stability control following a grid disturbance, support of quick communication protocol of IEC 61850 process layer, and clock synchronization with SNTP, IEEE 1588, or IRIG-B time code. It collects measured data from the local control layer and decides which information needs to be transmitted to the microgrid control center (MGCC), receives instructions from the MGCC, and transmits them to the local control layer. This requires a high data transmission capability and quality; therefore, long-distance and large-capacity communication technologies should be used. The PTN and IEC 61850-9-2 + GOOSE protocol can achieve flexible access, aggregation, and transmission of data from the local control layer.
   a. *PTN*: The PTN technology, with an IP core, can effectively fulfill the convergence and transmission of a larger amount of small-granule data services, and can best address transmission of the large amount and suddenly rising data services at the central control layer. In addition, PTN inherits the high protection and OAM capabilities of transmission equipment, and can provide system-level protection and monitoring management for data transmission.

      When using the PTN technology in a microgrid, a GE (for a small microgrid) or 10 GE (for a large microgrid) ring network is usually set up. The introduction of the PTN technology at the central control layer is mainly because of its advantages in flexibility, convergence of services on two layers, and statistical multiplexing of service transmission and access, to provide reliable, real-time, and flexible communication.
   b. *IEC 61850-9-2 + GOOSE communication protocol*: GOOSE is a real-time communication protocol that can realize tripping and closing of switchgear and operation control of protection equipment. It replaces the traditional hard wiring between intelligent electronic devices (IEDs), improving the speed, efficiency, and reliability of communication between logic nodes. It realizes true point-to-point communication. With this protocol, any IED can connect to another through Ethernet, receive data for the subscription server, and provide data for the push server. Network connection reduces equipment maintenance costs.

3. *Communication design of the distribution network dispatch layer*: The distribution network dispatch layer is expected for economic and optimal dispatch of the microgrid, forecast of outputs of DGs and loads, smooth control of power outputs of DGs, quick transfer control of master power source, quick energy balance control, and economic and optimal operation of the grid. This necessitates support of quick, real-time, and reliable communication technologies. To enhance the security, efficiency, and intelligence of the communication system of the microgrid, adapt it to a smart and distributed microgrid in the future, and meet demands of new services, an effective solution is to combine large-capacity and long-distance OTN and ASON technology for communication of the MGCC to realize seamless connection with the conventional distribution network and intelligent management of the microgrid.

Chapter 8

Earthing of a microgrid

Earthing is one of the most widely used electrical safety measures in daily industrial production, work, and life. Users are generally unfamiliar with earthing protection measures. Any problem in earthing of the distribution system may pose a threat to the safety of personnel and equipment. There are two kinds of earthing, namely, functional earthing and protective earthing. Functional earthing includes working earthing, logic earthing, signal earthing, and shield earthing. Protective earthing means connecting a part of the electrical equipment to the earth by an earthing device to ensure the safety of personnel and equipment. This chapter discusses protective earthing, which, by securely earthing the electrical equipment, ensures the safety of personnel and equipment.

8.1 SECURE EARTHING OF LOW-VOLTAGE DISTRIBUTION NETWORK

Providing system earthing and protective earthing through conductors can ensure safety and provide protection. System earthing means the earthing of a point (generally the neutral point) of the power source, and protective earthing means earthing of electrical equipment, electrical installation or system. Two kinds of conductors are used, namely, a neutral conductor (represented by N) and a protective conductor (represented by PE). The former is connected with the system neutral point and can transmit electricity. It is a neutral line and working line, also called "zero wire" in single-phase systems. The latter, commonly known as "earth wire," is used to earth the exposed conductive part of electrical equipment. Without the former, single-phase equipment cannot work normally; without the latter, equipment can be functional while its enclosure may be energized. The PE conductor can prevent electric shock.

As stated in GB 14050-2008 *Types and Safety Technical Requirements of System Earthing*, the earthing of 380/220 V AC low-voltage (LV) distribution systems is mainly in three forms, namely, TN, TT, and IT, and TN is

Microgrid Technology and Engineering Application

further divided into three forms, namely, TN-C, TN-S, and TN-C-S (a total of five forms).

The meanings of these letters are as follows: The first letter indicates the relation between the power source and the earth, specifically, T means that a point (usually the neutral point) of the power source is directly earthed, I means that the power source is isolated from the earth or a point of the power source is directly earthed via a high impedance; the second letter indicates the relation between the exposed conductive part of electrical installation and the earth, specifically, T means that the exposed conductive part is directly earthed (irrelevant to earthing of the power source), and N means that the exposed conductive part is connected to the earthed neutral point of the power source; the third and fourth letters indicate the relation between N conductor and PE conductor, specifically, C indicates the combination of the two, and S indicates separation of the two.

8.1.1 TN earthing system

TN earthing is a prevailing form in China. In a TN earthing system, the neutral point of the power source is directly earthed, and the exposed conductive part of power-consuming equipment is connected to the N conductor via PE conductor.

The advantage of TN earthing is that when a single-line-to-earth short circuit occurs due to contact of a phase line of electrical equipment with the enclosure or due to damage of insulation of the equipment, and the fault current rises, overcurrent protection will act to instantaneously cut off the power. Ideally, when such a fault occurs, the fuse on the source side will break and the LV circuit breaker will trip instantly, thus disconnecting the faulty equipment from the power source, shortening the duration of hazardous touch voltage, and improving security. In addition, TN earthing requires fewer materials and work time and is used in a wide range of applications. Currently, it is widely used in China. In the case of no neutral point, or when the neutral point is not led out, a phase on the secondary side of the transformer may be earthed instead. Depending on the relation between the N conductor and PE conductor, TN earthing can be divided into three forms, respectively, TN-C, TN-S, and TN-C-S.

1. TN-C earthing system
   Figure 8.1 shows the schematic diagram of a TN-C earthing system, in which the N conductor and PE conductor are combined as PEN conductor. Such a system is traditionally called a three-phase four-wire system.

   Its main advantages include reducing a conductor, plus simple installation and saving of costs.

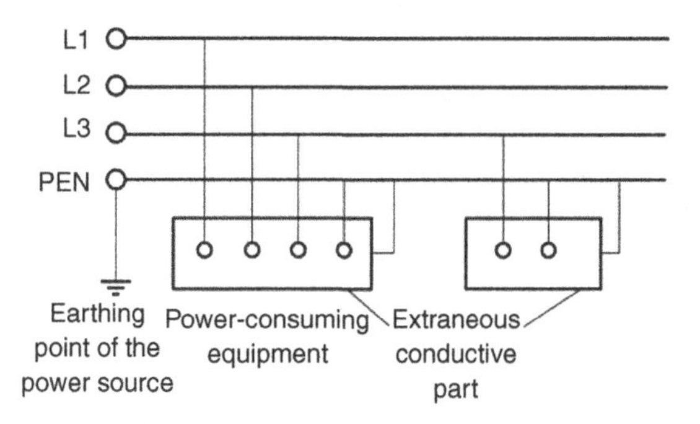

FIGURE 8.1 Schematic diagram of a TN-C earthing system.

However, it also has several disadvantages. When the PEN conductor is disconnected, an earthing fault will occur, the enclosure of the electrical equipment will be charged with a 220 V voltage, and thus pose hazards; as the PEN conductor cannot be disconnected, it is hard to make electrical isolation during maintenance; when a current flows through the PEN conductor, a voltage drop will occur, leaving a potential difference between the enclosure of the equipment connected with the conductor and the earth, which may interfere with the communication device or cause explosion in an explosive atmosphere.

Note: In the event of an earthing fault, the magnetic field of the current flowing through the PEN conductor affects correct operation of the residual current device (RCD), and poses hazards for preventing disconnection of the PE conductor. Therefore, repeated earthing shall be made, which, however, will easily cause displacement of the neutral point. For this reason, the PE conductor or repeated earthing system of loads should be connected to the PEN conductor on the power source side of RCD, as shown in Figure 8.2.

2. TN-S earthing system

Figure 8.3 shows the schematic diagram of a TN-S earthing system, in which the N conductor and PE conductor are separated. This system is traditionally called a three-phase five-wire system.

Its main advantages are high security and reliability. In the case of load unbalancing, there will be unbalanced current flowing through the N conductor while the PE conductor is not affected and has no potential to the earth. There is fault current flowing through the PE conductor only in contact with the enclosure, and normally, the exposed conductive part of the electrical equipment has zero potential to the earth; the switches at four levels can disconnect L1, L2, L3, and N conductors as necessary, thus making electrical isolation possible during maintenance.

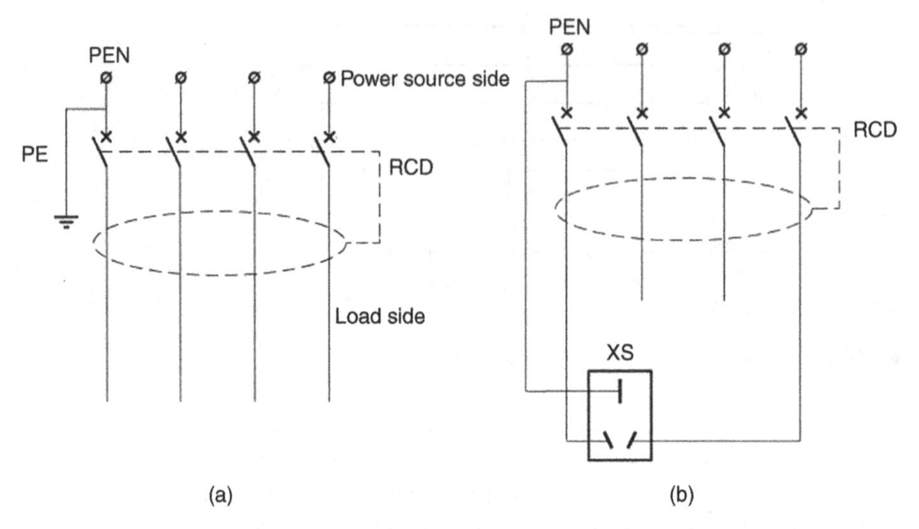

■ **FIGURE 8.2 RCD in a TN-C earthing system.** (a) Repeated earthing and (b) PE conductor.

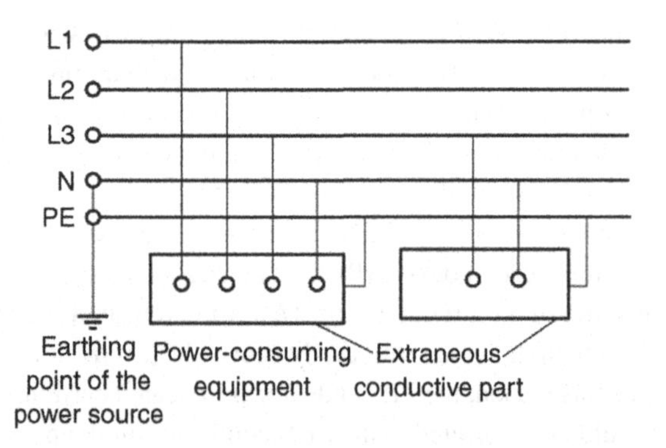

■ **FIGURE 8.3 Schematic diagram of a TN-S earthing system.**

However, a separate PE conductor needs to be erected all the way, requiring a high initial investment.

Note: The PE conductor must not be disconnected, but can be earthed repeatedly; the RCD may be used for preventing electrical shock and lightning strike, and the N conductor should in no case be earthed repeatedly.

3. TN-C-S earthing system

   Figure 8.4 shows the schematic diagram of a TN-C-S earthing system, in which the N conductor and PE conductor are combined (TN-C earthing) before the entry point into the building, and are separated (TN-S earthing) after the entry point into the building. Such a system is traditionally called a local three-phase five-wire system.

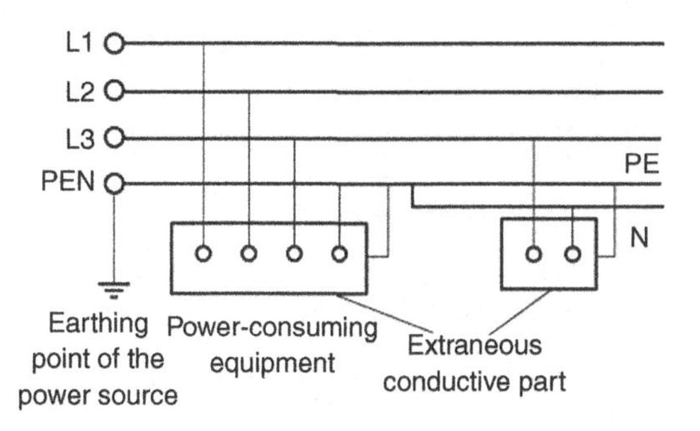

■ **FIGURE 8.4 Schematic diagram of a TN-C-S earthing system.**

For this system, it is unnecessary to erect a special PE conductor from the power source to the consumers' equipment. Although all conductive parts are interconnected at the entry point (i.e., main equipotential bonding, or MEB for short), and that the PEN conductor is split into separate PE conductor and N conductor after the entry point, no voltage drop appears in the PE conductor, and the entire electrical installation has zero potential to the earth. To sum up, such a system is free of the hazards that may arise in a TN-C earthing system and has a security level as a TN-S earthing system.

However, the voltage drop along the PEN conductor increases the voltage of the entire electrical installation to the earth. The amplitude of the voltage depends on the unbalanced current through the line and length of the line before combination. The more unbalanced the loads and longer the line before combination, the greater the displacement of the equipment enclosure-to-earth voltage. Therefore, the unbalanced current of loads is required to be within a certain range.

Note: The PE conductor shall be earthed repeatedly to prevent hazards resulting from disconnection of the conductor, which, however, may easily cause displacement of the neutral point.

Whether in a TN-C, TN-S, or TN-C-S earthing system, the fault voltage appearing on the PE conductor or PEN conductor will spread among various electrical installations energized by the same power source. Therefore, MEB shall be made to prevent any accident due to spreading of the fault voltage. For this reason, TN earthing should not be used in occasions without MEB such as street lights, construction sites, or agricultural applications.

8.1.2 TT earthing system

Figure 8.5 shows the schematic diagram of a TT earthing system, in which the neutral point of the power source is directly earthed, and the N conductor

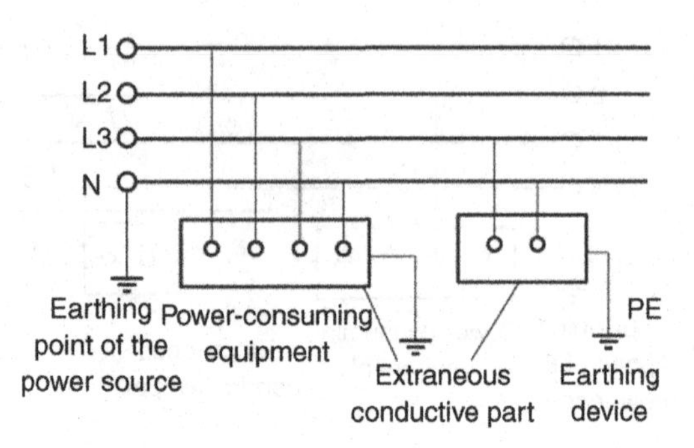

■ **FIGURE 8.5 Schematic diagram of a TT earthing system.**

is led out to constitute a three-phase four-wire system; the exposed conductive parts of power-consuming equipment are separately earthed via their respective PE conductors, thus realizing electrical isolation with system earthing.

The PE conductors of the exposed conductive parts of electrical installation are isolated from system earthing and independent of each other. Normally, the exposed conductive parts have a zero potential to the earth, and the fault voltage appearing on the power source side or various power-consuming equipment will not spread.

In the event of three-phase load unbalance after disconnection of the N conductor, potential drift of the neutral point will occur, causing voltage rise on a phase and thus burning of single-phase equipment; in the event of earthing fault, the fault current can return to the power source via the protective earthing resistor and system earthing resistor, and due to the resistance, the fault current is relatively small, thus failing to initiate overcurrent protection, and making the setting of overcurrent protection much more complicated.

Note: In the event of earthing fault, the fault current is relatively small due to the earthing resistance, and thus the LV automatic switch may fail to isolate the fault quickly, leaving a voltage on the enclosure of equipment. Therefore, an RCD is required for prevention against electric shock and lightning strike.

8.1.3 IT earthing system

Figure 8.6 shows the schematic diagram of an IT earthing system, in which the power source is not earthed or earthed via an impedance, the live parts on the power source side are insulated from the earth or earthed via a high impedance, the exposed conductive parts of electrical equipment are directly earthed, and the metallic enclosures of power-consuming equipment are

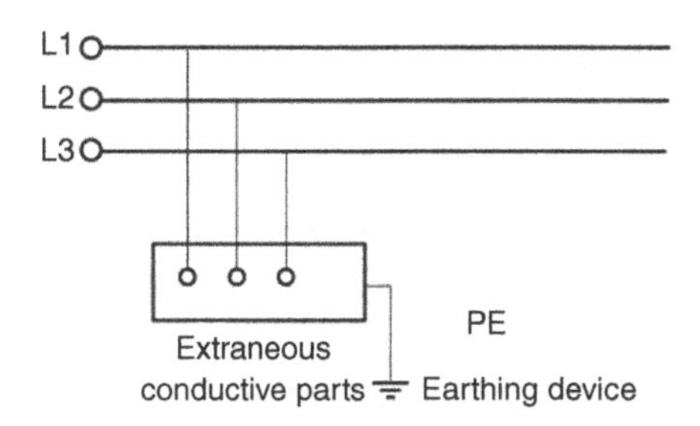

■ **FIGURE 8.6 Schematic diagram of an IT earthing system.**

directly earthed or connected to the earthing electrode of the power source via PE conductor.

It has the following advantages: in the case of a single earthing fault, due to the lack of a circuit for the fault current to return to the power source, the fault current is just a capacitive current to earth of the healthy phase and very small, and therefore, the fault voltage to the earth is very low and will not cause electric shock, explosion, or fire. Therefore, IT earthing is applicable to special occasions with high electrical hazards. Furthermore, in the case of a single earthing fault, it is not necessary to disconnect the power source to interrupt power, thus making IT earthing very suitable for electrical installations having a high requirement for uninterrupted power supply.

In this system, no N conductor is led out to provide 220 V power supply for single-phase equipment; in the case of single-phase-to-earth faults, the voltage of the other two phases will rise to a 380 V line voltage from a 220 V phase voltage, posing a greater threat to personal safety.

Note: Single-phase-to-earth fault is hard to detect; load unbalance will result in displacement of the neutral point. To detect single-phase-to-earth fault, an insulation monitoring device or single-phase earthing protection shall be provided to give audio and visual signals when a fault occurs to remind the operators to isolate the fault in time; otherwise, the fault will develop into two-phase-to-earth fault if earthing fault occurs in another phase, which will cause interruption of the power supply.

In view of the advantages and disadvantages of each individual earthing system, it is concluded that, in determining the earthing mode of an LV distribution system, TN-C earthing should be selected with due care; for a building with a substation, TN-S earthing is preferred, while for a building without a substation and with an LV power supply, TN-C-S earthing in conjunction with MEB should be employed; for an outdoor place without MEB, TT earthing should be employed in conjunction with RCD; for a place having a high requirement on power reliability and with inflammable and explosive materials, IT earthing should be employed.

8.2 SYSTEM EARTHING OF A MICROGRID

8.2.1 Requirements for earthing of microgrid

Grid-connection of distributed resources (DRs) should follow the requirements in Q/GDW 480-2010 *Technical Rules for Distributed Resources Connected to Power Grid*, specifically, the earthing mode of DRs shall be consistent with that of the grid, and can ensure personnel and equipment safety and protection coordination. IEEE 1547 *IEEE Standard for Interconnecting Distributed Resources to Electric Power Systems* specifies the basic principles for earthing of the microgrid as follows: the earthing mode of DRs should not cause overvoltage of equipment within the microgrid and affect coordination of protections. For distributed generation (DG) that is connected to the distribution network not as a microgrid, it is usually connected with the neutral point not earthed.

Depending on the earthing mode of the LV distribution system, the microgrid can adopt three earthing modes: (1) the neutral point of DG is not earthed and the N conductor is not led out, (2) the neutral point of DG is not earthed but the N conductor is led out, (3) and the neutral point of DG is earthed. The earthing mode of the microgrid should be properly determined for different earthing systems.

8.2.2 Earthing mode of DG in a TN or TT earthing system

If, in a TN or TT earthing system, the neutral point of DG is not earthed and the N conductor is not led out, a 220 V voltage cannot be provided for single-phase loads in islanded operation, as shown in Figure 8.7a. While if the neutral point of DG is earthed, as shown in Figure 8.7c, in grid-connected operation, a small proportion of the three-phase unbalanced current will return to the power source not via the N conductor as it is supposed to but via an unexpected circuit. This proportion of current is called stray current. This stray current will be high if there are multiple DGs or if the transformer and DG are located in the same building. This may result in the following accidents:

1. *Fire*: The stray current may ignite combustible materials if the circuit is poorly conductive.
2. *Corrosion of underground metallic parts*: If returning to the power source through the earth circuit, the stray current may cause corrosion of underground metallic pipelines or foundation reinforcement.

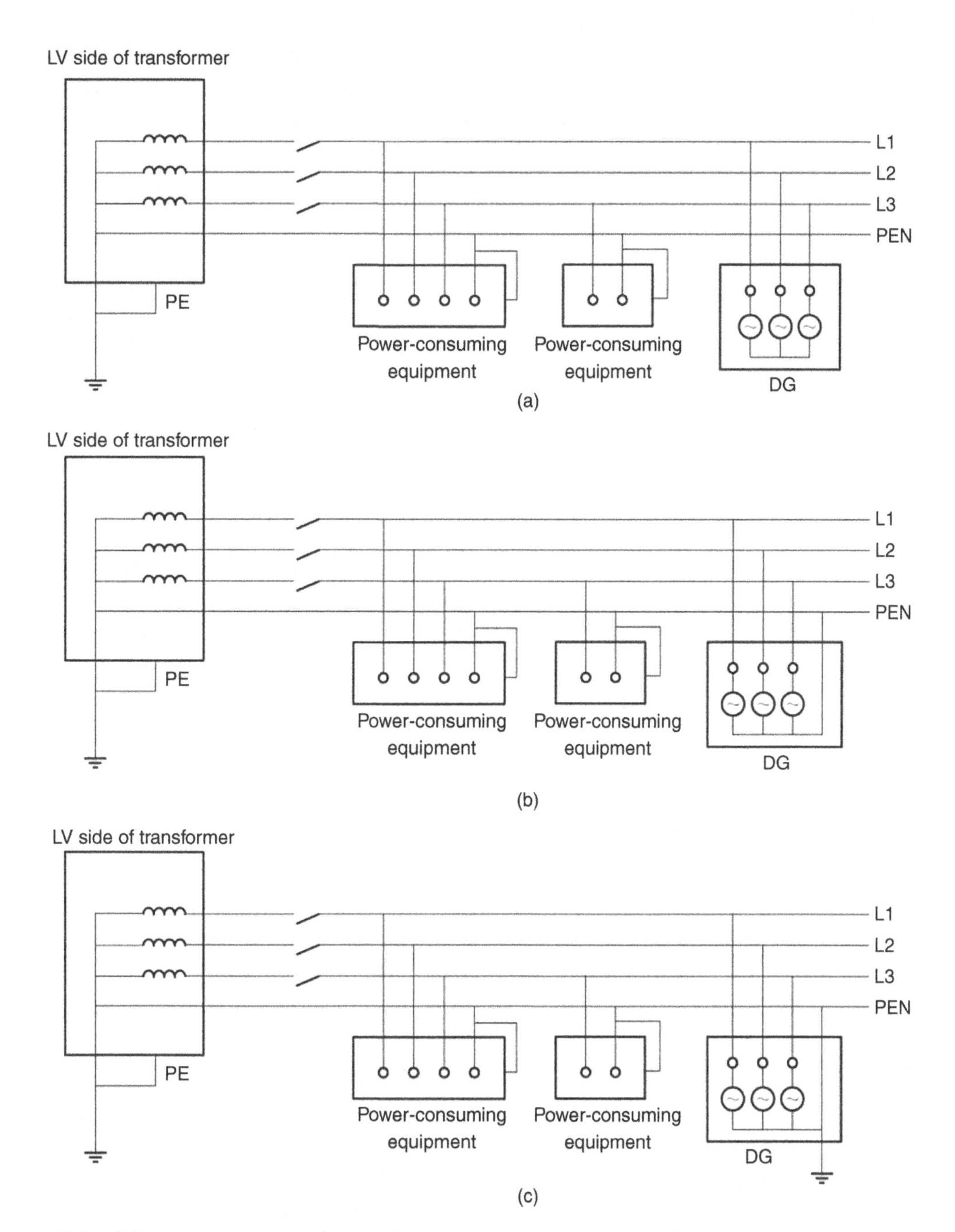

■ **FIGURE 8.7 Earthing mode of DG in a TN or TT earthing system.** (a) N conductor not led out, (b) N conductor led out, and (c) neutral point earthed.

3. *Interference of magnetic fields*: The power distribution circuit and the stray current-to-power source circuit may form a closed ring. The magnetic fields within the ring may cause interference to the highly sensitive communication equipment within the ring.

IEC 60364 – 1: 2005 *Low-voltage Electrical Installations–Part 1: Fundamental Principles, Assessment of General Characteristics, Definitions*, IEC 60364 – 4-444 *Electrical Installations of Buildings–Part 4: Protection for Safety Section 444: Protection against Electro-Magnetic Interferences (EMI) in Installations of Buildings* and China's standard GB/T 16895.1 – 2008 *Low-voltage Electrical Installations–Part 1: Fundamental Principles, Assessment of General Characteristics, Definitions* all stipulate that a multisource system shall be earthed only at one point, to avoid any unwanted stray current circuit due to improper earthing (see Figure 8.8). The following lists the requirements:

1. The neutral points of transformers or the star points of generators should not be directly connected to the earth.
2. The conductor interconnecting the neutral points of transformers or star points of generators should be insulated and not be connected with power-consuming equipment.

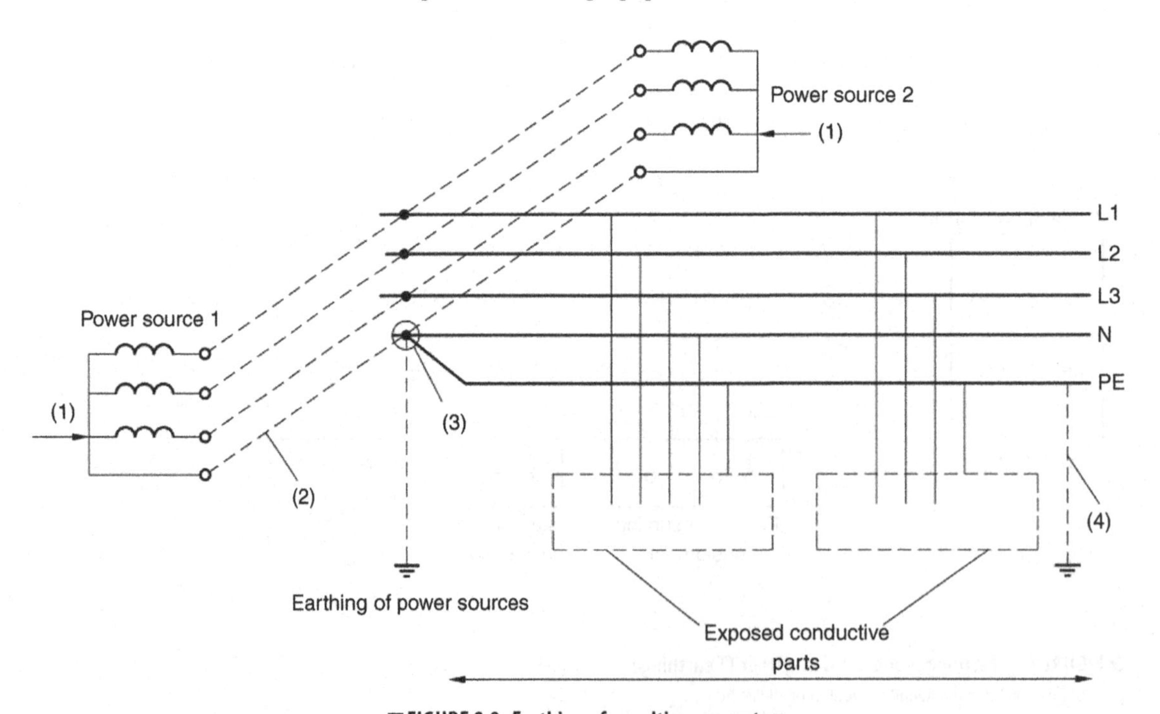

■ **FIGURE 8.8 Earthing of a multisource system.**

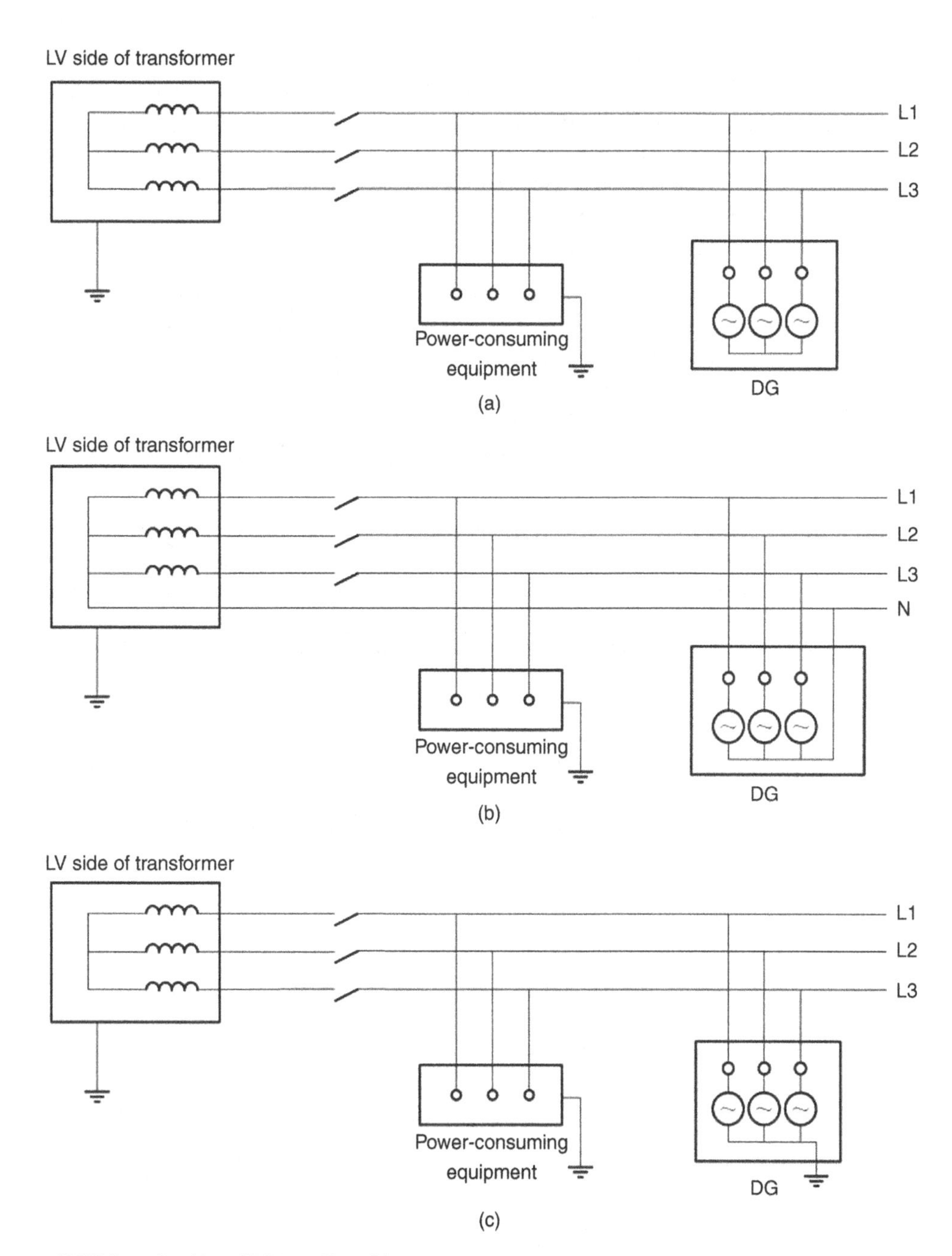

FIGURE 8.9 Earthing of DG in an IT earthing system. (a) N conductor not led out, (b) N conductor led out, and (c) neutral point earthed.

3. The conductor interconnecting the neutral points of power sources shall be connected with the PE conductor only at one point and within the main distribution panel.
4. The PE conductor connecting to the electrical installation may be provided with additional earthing device.
5. No extension of the system should affect the normal functions of protection measures.

In conclusion, in a TN or TT earthing system, the DGs may be earthed in such a way that the neutral point is not directly earthed, and the N conductor is led out and earthed at one point within the LV main distribution panel, as shown in Figure 8.7b.

8.2.3 Earthing mode of DG in an IT earthing system

In an IT earthing system, all live parts are isolated from the earth, and the exposed conductive parts of the electrical installation are separately earthed. The integration of DGs should not change the original earthing mode. If in an IT earthing system, the neutral point of DG is earthed as shown in Figure 8.9c, IT earthing will become TN or TT earthing, which is worse than IT earthing in terms of power reliability and security.

In an IT earthing system, the N conductor is generally not led out, and in this case, the N conductor of DG may not be led out either, as shown in Figure 8.9a; while if the N conductor in an IT earthing system is led out, the N conductor of DG may be also led out, as shown in Figure 8.9b, to ensure safety and reliability in grid-connected operation and islanded operation.

Chapter 9

Harmonic control of the microgrid

The harmonics in microgrid come from two sources: background harmonics produced from the power system connected with the microgrid, and harmonics produced by the nonlinear loads and equipment (including inverters) of the microgrid. Effective harmonic control can improve power quality and ensure normal operation of loads.

In a microgrid, most distributed generation (DG) or power systems are connected to the grid through power electronic converters, such as grid tie inverter and power converter system (PCS) for photovoltaic (PV) power, and wind power. These devices are harmonic sources. With the advance of technologies, these devices will have more functions. It has become a trend for grid tie inverters to work also for power quality control.

9.1 HARMONIC CONTROL TECHNOLOGIES

9.1.1 Passive filtering of grid tie inverter

A passive circuit can be used to remove harmonics generated by the grid tie inverter, mostly harmonics at around the switching frequency of the inverter, or high-order harmonics of an integer multiple of the switching frequency. Let the switching frequency of the inverter be f_k, then the frequency of the harmonics generated by the inverter is $nf_k + 1$. Figure 9.1 shows the topology of the passive filtering circuit, which can only remove the high-order harmonics generated by the inverter, but not those generated by loads.

9.1.2 Active filtering of grid tie inverter

With proper control means, a grid tie inverter can also act as an active power filter. This requires the sinusoidal pulse width modulation (SPWM) technology to generate a desired harmonic current to be superimposed on the sinusoidal voltage generated on the AC side of the inverter, provided that the said harmonic current is equal to but in opposite phase with the harmonic

Microgrid Technology and Engineering Application

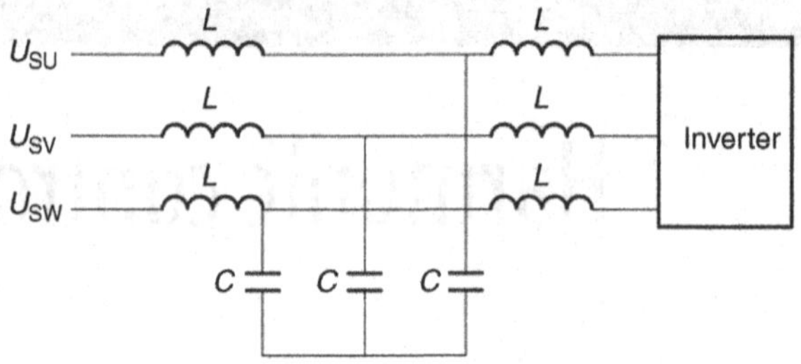

■ FIGURE 9.1 Passive filtering circuit of grid tie inverter.

current generated by loads of the grid. Such an inverter is called an active filtering grid tie inverter.

9.1.3 Separately configured filtering technologies

If the grid tie inverter is properly controlled such that the iron-core reactor series-connected with the inverter is free from magnetic saturation, harmonics injected to the grid from the inverter can be ignored. However, since the loads also generate harmonics, the grid still suffers excessive harmonics under some circumstances. As such, the control system of the microgrid should also be equipped with active or passive power filters sometimes.

A low-voltage (LV) *LC* single-tuned filter can be used as the passive filtering circuit for a microgrid. The single-tuned filter is designed with a certain margin considering the effects of quality factor during resonance. Given the relatively high fundamental current in an LV system, the passive high-pass filter is not recommended.

An active power filter composed of power electronic devices may be used as the active filtering circuit for microgrid. As the harmonic current occurring in the microgrid is relatively low, an active power filter with a limited capacity can provide sufficient harmonic control.

9.2 PASSIVE FILTERING TECHNOLOGIES

9.2.1 Basic principles of passive filter

A passive filter is an appropriate combination of filter capacitor, reactor, and resistor (or using the resistance provided by the reactor). By connecting it in parallel with the harmonic source, it can both filter harmonics and provide reactive power compensation. It is also known as *LC* filter as it is composed of inductors (represented by the letter *L*) and capacitors (represented by the letter *C*). *LC* filters can be further divided into single-tuned filter, double-tuned filter, and high-pass filter, with the first and third types often used.

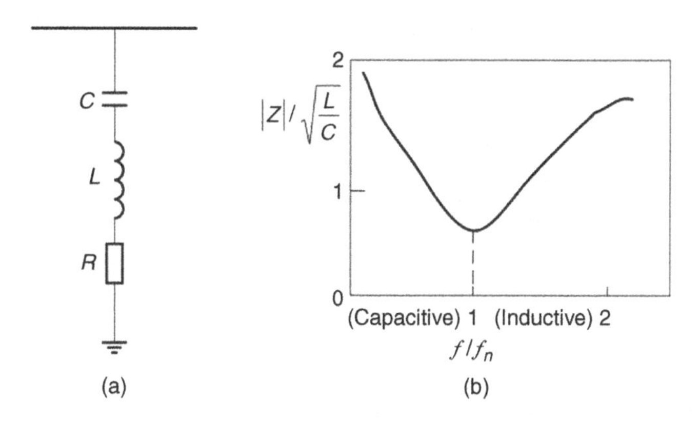

FIGURE 9.2 Circuit diagram and impedance-frequency characteristic of single-tuned filter. (a) Circuit diagram and (b) impedance-frequency characteristic.

9.2.1.1 ***Single-tuned filter***

Figure 9.2a shows the circuit diagram of a single-tuned filter. The impedance of the filter with respect to *n*th harmonic $\omega_n = n\omega_s$ is

$$Z_{\text{fn}} = R_{\text{fn}} + j\left(n\omega_s L - \frac{1}{n\omega_s C}\right) \tag{9.1}$$

where fn refers to the single-tuned filter for *n*th harmonics.

The impedance-frequency curve of the filter is shown in Figure 9.2b.

A single-tuned filter works based on the *LC* resonant circuit, and the harmonic order *n* at the point of resonance is given by:

$$n = \frac{1}{\omega_s \sqrt{LC}} \tag{9.2}$$

At the point of resonance, $Z_{\text{fn}} = R_{\text{fn}}$. Since R_{fn} is very small, most of the *n*th harmonic current flows to the filter while only a small portion flows to the grid. While for harmonics of other orders, $Z_{\text{fn}} \neq R_{\text{fn}}$, and only a minor portion of current flows to the filter. Simply put, as long as the harmonic order at the point of resonance is set to the order of harmonics to be removed, most of harmonic current of this order will flow to the filter and thus effectively be removed.

9.2.1.2 ***High-pass filter***

High-pass filter, also known as damped filter, comes in four types, respectively first-order type, second-order type, third-order type, and C-type, as shown in Figure 9.3.

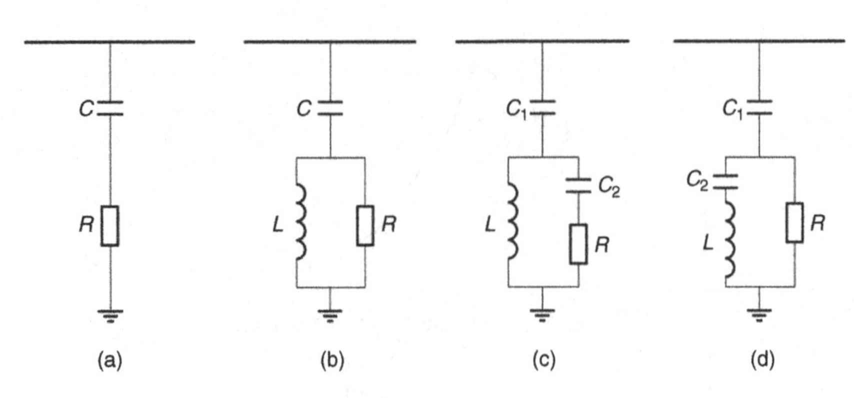

FIGURE 9.3 Circuit diagrams of high-pass filter. (a) First-order, (b) second-order, (c) third-order, and (d) C-type.

The first-order high-pass filter requires a capacitor with a very high capacity and thus causes very high fundamental losses, and therefore, is seldom used.

The second-order high-pass filter has the best performance, but causes higher fundamental losses compared with the third-order type.

The third-order high-pass filter has an additional capacitor C_2 compared with the second-order type. It has a capacity much lower than that of C_1, thus increasing the impedance of the filter to the fundamental frequency, and significantly reducing fundamental losses. This is a major advantage of the third-order high-pass filter.

The performance of C-type high-pass filter falls between the third-order type and the second-order type. C_2 and L are tuned at the fundamental frequency, thus significantly reducing fundamental losses. But the disadvantage is that it is sensitive to detuning of fundamental frequency and parameter drift of devices.

Among the above four types of high-pass filters, the most commonly used is second-order type at present. Its impedance is given by:

$$Z_n = \frac{1}{jn\omega_s C} + \left(\frac{1}{R} + \frac{1}{jn\omega_s C}\right)^{-1} \tag{9.3}$$

9.2.1.3 Double-tuned filter

In addition to single-tuned filter and high-pass filter, double-tuned filters are also used. Figure 9.4 shows the circuit diagram of a double-tuned filter. It is equivalent to two parallel-connected single-tuned filters as it has two resonance frequencies and absorbs harmonics at these two frequencies.

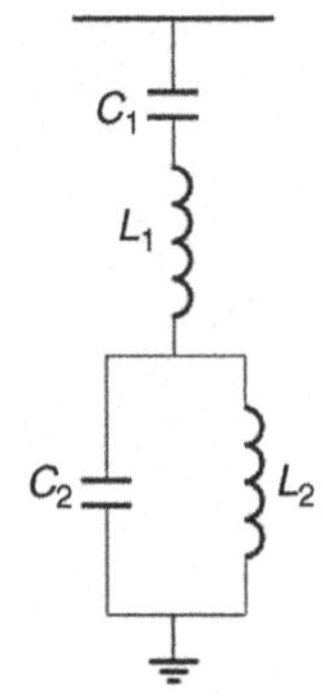

FIGURE 9.4 Circuit diagram of double-tuned filter.

Compared with two single-tuned filters, a double-tuned filter causes lower fundamental losses and uses only one inductor (L_1) to withstand all the surge voltage. In normal operation, the fundamental-frequency impedance in the series circuit is much higher than that in the parallel circuit; therefore, the power-frequency voltage occurring in the latter circuit is much lower than that in the former circuit. Moreover, only the reactive component of the harmonic current flows through the parallel circuit as the capacity of C_2 is generally pretty small. On the other hand, the double-tuned filter has a rather complicated structure and is difficult to tune. Due to the low cost, double-tuned filters have been applied in some HVDC projects worldwide in recent years.

9.2.2 Design of passive filter

1. *Measurement and analysis of harmonic sources*: Prior measurement and analysis of the power system and its loads is required to determine the harmonic voltage and current of various orders injected to the system from the harmonic sources, the harmonic distortion rate, the harmonic impedance and reference capacity of the system, and measure the positive and negative frequency offsets and background harmonics of the system (the effects of background harmonics may be considered as 110% of various orders of harmonics generated by the harmonic sources); analyze the composition and operation modes of loads; and identify the causes for harmonics, the operating time and changes of loads generating harmonics, and reactive power compensation.
2. *Determination of filter design*: Filter design involves determination of filter type and the corresponding harmonic order. Single-tuned filters are mainly used to remove high-content characteristic harmonics while high-pass filters remove higher-order harmonics excluding characteristic harmonics.
3. *Calculation of parameters of single-tuned filter*: The parameters include the impedance, withstand voltage, and capacity of the reactor and capacitor comprising the filter.

9.3 ACTIVE FILTERING TECHNOLOGIES

Figure 9.5 shows the circuit of a three-phase active power filter, where L_R is the filter reactor, which also serves to convert the voltage to current, C_R is a high-frequency filter capacitor, VT_1, VT_2, VT_3, VT_4, VT_5, and VT_6 are IGBT components, VD1, VD_2, VD_3, VD_4, VD_5, and VD_6 are fast recovery diodes, and C is an electrolytic capacitor. The filter is generally parallel connected with nonlinear loads and injects to the grid a harmonic current equal to but in opposite phase with the harmonic current generated by the nonlinear

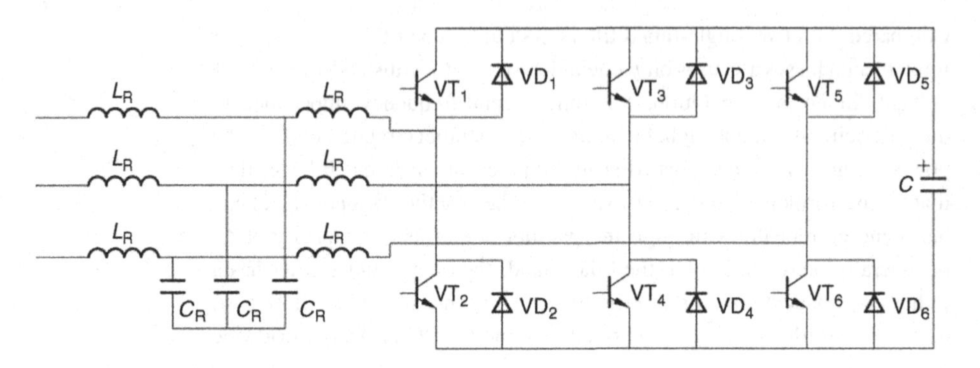

■ **FIGURE 9.5 Circuit diagram of three-phase active power filter.**

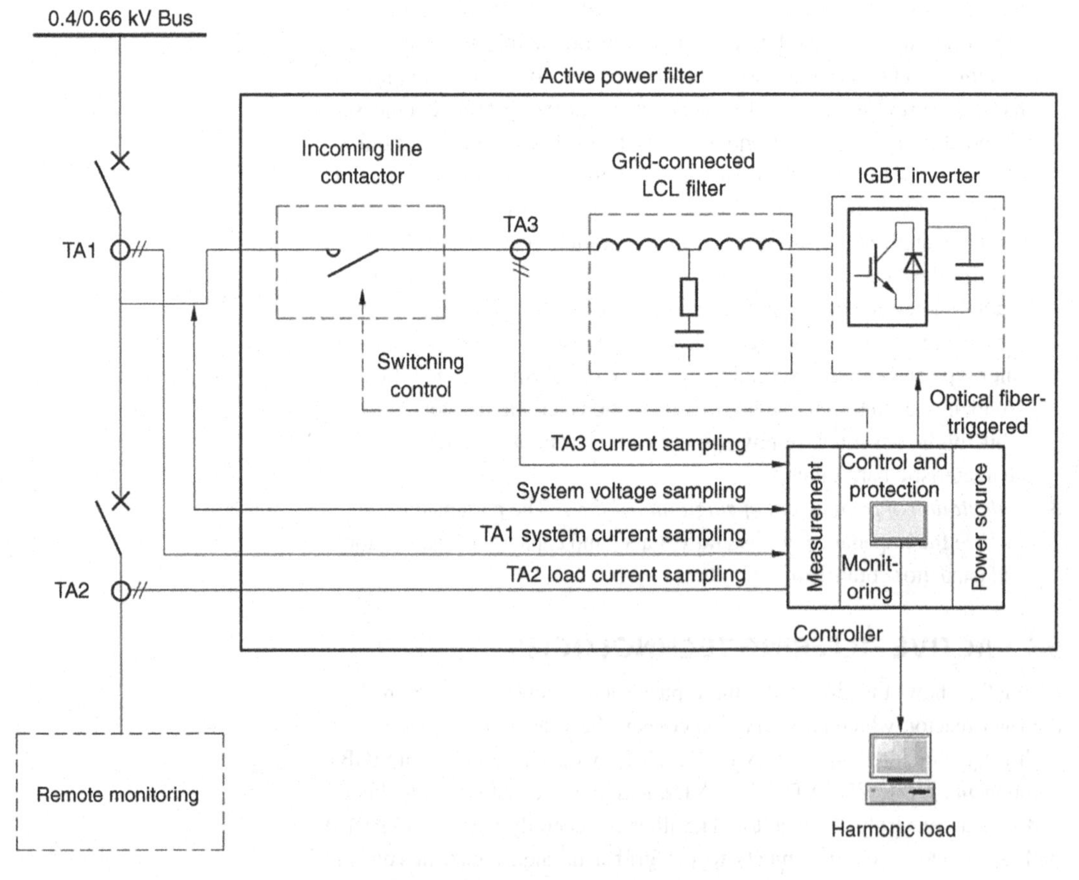

■ **FIGURE 9.6 Block diagram of system configuration.**

loads, thus making the total harmonic current approach zero and effectively eliminating the effects of harmonics.

The active power filter for a microgrid is a voltage source inverter composed of power electronic devices. It operates as a controlled current source and its output current, having the same frequency and phase as the system voltage, flows to the system through the connecting reactor. By detecting the harmonic current induced by loads in real time, the filter is controlled to inject to the power system a current with the same waveform and opposite direction as the load induced harmonic current, for the purpose of harmonic compensation. An active power filter mainly has the following functions:

1. Dynamically suppress the harmonic current flowing to the grid, to reduce the THD content of the current;
2. Improve the power quality of the distribution network;
3. Compensate reactive power for loads, to increase the power factor and save energy;
4. Prevent occurrence of *LC* resonance in the grid, improve security of the distribution network, and protect reactive compensation devices;
5. Reduce feeder losses and improve the efficiency of the distribution network.

An active power filter is mainly composed of incoming line contactor, filter, IGBT inverter, and control and protection, as shown in Figure 9.6. The control system of the active power filter detects the current on the load side and the system voltage at the connection point in real time, calculates the current required for harmonic compensation, and controls the three-phase IGBT inverters to give current tracking and compensation orders to suppress harmonics on the load side. Besides, the filter is also capable of protection against AC overvoltage/undervoltage, output overcurrent, DC overvoltage/undervoltage, short circuit of inverter, and overheating of inverter.

Chapter 10

Related standards and specifications

The rapid development of the microgrid technology and various distributed resources (DRs), such as wind, photovoltaics (PV), and biomass, is changing the traditional radial pattern of power systems, and poses challenges to the security and stability of the grid. Unified standards and specifications for power quality, active/reactive power control, security, protection, and communication of DRs must be available to ensure orderly development of DRs and security and reliability of the power system.

10.1 RELATED INTERNATIONAL STANDARDS AND SPECIFICATIONS

The microgrid is an emerging type of power network. Up to now, there has been no unified, systematic standard or specification officially issued in China for the microgrid, but some international standards have incorporated related specifications.

Currently, the IEEE 1547 *IEEE Standard for Interconnecting Distributed Resources with Electric Power Systems*, IEC TS 62257 *Recommendations for Small Renewable Energy and Hybrid Systems for Rural Electrification*, and DG interconnection and power quality standards in some countries form the framework of international standards on distributed generation. Among them, the IEEE 1547 is the most widely accepted standard. The provisions pertaining to the microgrid in these standards provide reference for the preparation of microgrid-specific standards in the future. However, only IEEE 1547.4 has incorporated such provisions, which are just a draft version and have not been officially issued.

10.1.1 Related IEEE standards

An early international standard on DRs is IEEE 446-1995 *Recommended Practice for Emergency and Standby Power Systems for Industrial and Commercial Applications*, which states how emergency and standby power

Microgrid Technology and Engineering Application

sources are installed and applied, and that users can power loads with DRs, and such generators are mainly for reliable power supply in an emergency and not interconnected to the grid.

The IEEE 1547 was officially published by the IEEE in 2003, and established as a national standard of the United States. It lays down the basic requirements for all aspects on interconnection of DRs with an aggregate capacity of 10 MVA or less, including power quality, system reliability, system protection, communication, security criteria, and metering.

The IEEE 1547 standard is not inclusive of interconnection of all DRs; it has some limitations. Specifically, it is applicable only to inverter-based DRs with an aggregate capacity of 10 MVA or less, and installed on a 60 Hz primary or secondary distribution system; it does not deal with DR self-protection and power system planning and design. As such, the IEEE 1547 has been gradually expanded to a series (as listed next), in order to standardize the interconnection of DRs, and reduce the costs for construction and operation of infrastructure related to renewable energy:

1. IEEE 1547.1 *Standard for Conformance Test Procedures for Equipment Interconnecting Distributed Resources with Electric Power Systems*. This standard, released in 2005, specifies the design, production, commissioning, and periodic connection test of DRs to determine whether DRs are suitable for interconnection.
2. IEEE 1547.2 *Application Guide for IEEE 1547 Standard for Interconnecting Distributed Resources with Electric Power Systems*. This standard provides technical background and application details to support understanding of IEEE 1547.
3. IEEE 1547.3 *IEEE Guide for Monitoring, Information Exchange, and Control of Distributed Resources Interconnected with Electric Power Systems*. This standard permits intentional islanding of a part of the distribution network to improve power reliability.
4. IEEE 1547.4 *Draft Guide for Design, Operation, and Integration of Distributed Resource Island Systems with Electric Power Systems*. This guide provides alternative approaches and good practices for design, operation, and integration of DR island systems, including the ability to separate from and reconnect to the grid.
5. IEEE 1547.5 *Draft Technical Guidelines for Interconnection of Electric Power Sources Greater than 10 MVA to the Power Transmission Grid*. This standard provides requirements for the design, construction, commissioning, acceptance, test, maintenance, and performance of DRs greater than 10 MVA interconnected to the transmission network.
6. IEEE 1547.6 *Draft Recommended Practice for Interconnecting Distributed Resources with Electric Power Systems Distribution*

Secondary Networks. This standard provides guidance for interconnecting DRs with electric power systems distribution secondary networks. It was sponsored by the IEEE Standards Coordinating Committee 21 on Fuel Cells, Photovoltaics, Dispersed Generation, and Energy Storage.

7. IEEE 1547.7 *Draft Guide to Conducting Distribution Impact Studies for Distributed Resource Interconnection*. This guide gives an engineering study method for analyzing the potential impacts of DR interconnection on regional distribution systems.
8. IEEE 1547.8 *Draft Recommended Practice for Establishing Methods and Procedures that Provide Supplemental Support for Implementation Strategies for Expanded Use of IEEE 1547*. This standard, providing more flexible design methods and procedures, expands the use of IEEE 1547.

Among them, the IEEE 1547.4 details the design and operation of the microgrid in grid-connected mode and islanded mode, the main considerations in planning and operating the microgrid, and discusses the strategies for grid-connected and islanded operation, thus providing reference for the preparation of microgrid-specific standards in the future.

The IEEE 519-1992 regulates that the installation of DG shall in no case cause voltage flicker beyond the limit, and defines the voltage flicker limit curve. The IEEE P1547 and IEEE 929-2000 are completely compatible in establishing standards for harmonics. The total harmonic distortion (THD) shall not exceed 5% of the rating at a voltage frequency of 60 Hz, or for any DG, the THD at the PCC shall not exceed 3% of the rating.

DGs are mostly connected to distribution systems at a low- or medium-voltage level. Currently, no country has specified the maximum voltage that DGs can connect to, and different standards give different voltage limits.

Most standards require that the voltage fluctuation at the PCC caused by installation of DG shall not be more than ±5% of the rated voltage. Almost no standard permits voltage control by DGs.

10.1.2 **Related IEC standards**

The main IEC standard relevant to the microgrid is IEC TS 62257 *Recommendations for Small Renewable Energy and Hybrid Systems for Rural Electrification*, which gives instructions on siting, equipment sizing, system design, and management of rural electrification projects. It applies to renewable energy and hybrid systems with AC voltage below 500 V and DC voltage below 50 V. It consists of the following nine sub-standards:

1. IEC TS 62257-1-2003 *Recommendations for Small Renewable Energy and Hybrid Systems for Rural Electrification-Part 1: General Introduction to Rural Electrification*

2. IEC TS 62257-2-2004 *Recommendations for Small Renewable Energy and Hybrid Systems for Rural Electrification-Part 2: From Requirements to a Range of Electrification Systems*
3. IEC TS 62257-3-2004 *Recommendations for Small Renewable Energy and Hybrid Systems for Rural Electrification-Part 3: Project Development and Management*
4. IEC TS 62257-4-2005 *Recommendations for Small Renewable Energy and Hybrid Systems for Rural Electrification-Part 4: System Selection and Design*
5. IEC TS 62257-5-2005 *Recommendations for Small Renewable Energy and Hybrid Systems for Rural Electrification-Part 5: Protection Against Electrical Hazards*
6. IEC TS 62257-6-2005 *Recommendations for Small Renewable Energy and Hybrid Systems for Rural Electrification-Part 6: Acceptance, Operation, Maintenance and Replacement*
7. IEC TS 62257-7-2008 *Recommendations for Small Renewable Energy and Hybrid Systems for Rural Electrification-Part 7: Generators*
8. IEC TS 62257-8-1-2007 *Recommendations for Small Renewable Energy and Hybrid Systems for Rural Electrification-Part 8-1: Selection of Batteries and Battery Management Systems for Stand-alone Electrification Systems-Specific Case of Automotive Flooded Lead-acid Batteries Available in Developing Countries*
9. IEC TS 62257-9-1-2008 *Recommendations for Small Renewable Energy and Hybrid Systems for Rural Electrification-Part 9-1: Micropower Systems*; IEC TS 62257-9-2-2008 *Recommendations for Small Renewable Energy and Hybrid Systems for Rural Electrification-Part 9-2: Microgrids*

Among these standards, IEC TS 62257-9-2 *Recommendations for Small Renewable Energy and Hybrid Systems for Rural Electrification-Part 9-2: Microgrids* provides special requirements for the design and implementation of dispersed rural microgrids and procedures to ensure personal and property safety, and detailed instructions and technical requirements on the limits, composition, voltage drops, protection against electric shocks, protection against overcurrents, selection and erection of equipment, and verification and acceptance of microgrids. It applies to low-voltage (LV) AC microgrids, three-phase or single-phase, with a capacity of 100 kVA or less.

10.1.3 Other related standards

Almost all countries in Europe have established their own requirements on the interconnection of DRs. The United Kingdom, France, Germany, and Belgium have developed standards, but they do not cover all DG technologies. The standards of the United Kingdom and Canada mainly focus on the requirements and specifications for connection of DGs, response of DGs to abnormality of distribution networks, and protection configurations, including islanding protection, power quality, and operation, security and tests of DGs.

Through discussions, the European Committee for Electrotechnical Standardization, CENELEC, issued *Requirements for the Connection of Micro-generators in Parallel with Public Low-voltage Distribution Networks (draft)*. The United Kingdom's standard ER G75/1 *Recommendations for the Connection of Embedded Generating Plant to Public Distribution Systems above 20 kV or with Outputs over 5 MW* specifies the respective responsibilities of utilities and power producers, defines the basic operating parameters, including voltage, frequency, grid structure, short-circuit current, and relay protection, analyzes the security, reliability, and stability of the electric power system, and considers islanded operation, making it a comprehensive standard on the grid-connection of a microgrid.

Canada established its interim codes on the development of a microgrid in July 2003, which focus on inverter-based microsources rated below 600 V, and include two interconnection standards, C22.2 No. 257 *Standard for Interconnection of Inverter-based Microsources with the Distribution Network* and C22.3 No. 9 *Standard for Interconnection between Distributed Electric Power Systems*. The former sets forth requirements for secure interconnection of inverter-based DRs with distribution networks below 0.6 kV, and the latter applies to DRs interconnected to a distribution network below 50 kV and with an aggregate capacity up to 10 MW and provides requirements and criteria for the performance, operation, test, and security of the DRs.

The Business Council of Australia formulated a renewable energy development guide titled *Guide to Connection of Microsources to the National Electric Power Market* in September 2003. This guide, in light of the Electricity Industry Act, provides guidance for application of microsources and briefly describes the process of and requirements for connecting microsources to the grid. It is applicable to microsources with a capacity up to 100 kW.

The Ministry of Economy, Trade, and Industry of Japan issued the *Guide to Interconnection of Distributed Resources* in May 1986, legalizing the interconnection of DRs to the grid. Later, the Ministry issued EAG 970-1993 *Technical Recommendations for Connection of Distributed Generation to the Grid* in 1994, amended the *Electric Power Law* in December 1995, and further revised the *Guide to Interconnection of Distributed Resources*, permitting the DR operators to sell the surplus energy to utilities and requiring the utilities to provide reserve for DR operators.

Most international interconnection standards and requirements can be classified into general requirements and specifications, security and protection requirements, and power quality requirements. Most interconnection standards are based on some common principles, including that the installation of DRs should not put other customers, the public or operators in danger, or affect the coordination of mechanical and electrical protections, reduce the reliability or limit the capacity of the electric power system; DRs should be equipped with protection devices and manual disconnection devices; only when the phase sequence, magnitude, phase, and frequency of the voltage of DRs are within the normal range can the DRs be connected with the electric power system and produce limited power.

10.2 RELATED STANDARDS AND SPECIFICATIONS IN CHINA

China's studies on DRs and the microgrid are still at an infant stage, and no uniform and systematic standards or specifications on the microgrid are available. At present, China's standards relevant to the microgrid mainly include power quality standards, interconnection standards for PV and wind power systems, and related distribution network standards.

China's power quality standards, mainly including the following, may be directly referenced or modified based on the specific conditions of the microgrid for interconnection of the microgrid to a utility grid:

1. GB/T 12325-2008 Power Quality – Deviation of Supply Voltage
2. GB/T 12326-2008 Power Quality – Voltage Fluctuation and Flicker
3. GB/T 14549-2003 Quality of Electric Energy Supply – Harmonics in Public Supply Network
4. GB/T 15543-2008 Power Quality – Three-Phase Voltage Unbalance
5. GB/T 15945-2008 Power Quality – Frequency Deviation for Power System

China has also formulated or amended standards on interconnection of wind and PV power systems, including GB/Z 19963-2005 *Technical Rule for Connecting Wind Farm to Power System*, Q/GDW 392-2009 *Technical Rule for Connecting Wind Farm into Power Grid*, GB/T 19939-2005 *Technical Requirements for Grid Connection of PV System*, GB/Z 19964-2005 *Technical Rule for Connecting PV Power Station to Electric Power System*, and Q/GDW 147-2010 *Technical Rule for Connecting PV Power Station to Power Grid (Interim)*. But for grid connection of other types of DRs, including biomass (agricultural biomass, forest biomass, methane, and waste), natural gas, coalbed methane, waste gas, industrial heat recovery, industrial excess pressure, geothermal energy, and ocean energy, as well as batteries and fuel cells, no uniform standards are available, thus making it difficult to determine the technical indices for interconnection of such DRs.

In November 25, 2011, as required in the *Notice of the Plan on Preparation and Revision of Energy Sector Standards (second batch) in 2011* (GNKJ (2011) No. 252) issued by the National Energy Administration of China, the China Electric Power Research Institute organized parties of interest to draft industry standards including *Test Technical Specifications for Equipment Connecting Distributed Resources with Power Grid*, *Specifications for Connecting Distributed Resources with Power Grid*, *Technical Specifications for Connecting Energy Storage System to Power Grid*, and *Operation and Control Specification for Electric Energy Storage System Interconnecting with Power Grid*.

With a view to providing guidance for the planning, design, and operation of various DRs for interconnection, and mitigating their impacts on the power system and ensuring the power quality, security, and reliability of the LV power system, the SGCC, taking into account the characteristics of various DRs, structure of grid at 35 kV or below, and requirements of grid operation on power sources, formulated the standard Q/GDW 480-2010 *Technical Rule for Distributed Resources Connected to Power Grid*, which provides specific technical requirements for grid connection of DRs, power quality, power control and voltage regulation, voltage, current and frequency response, security, relay protection and security automatic equipment, communication and information, energy metering, and grid connection tests.

10.3 DEVELOPMENT TREND OF MICROGRID STANDARDS

Up to now, there has been no official standard on the microgrid in China and around the world. Most microgrids are connected to a medium-voltage or LV distribution network, and thus mainly affect the planning and operation of the distribution network. Existing standards relevant to the microgrid mainly specify the planning, operation reliability and power quality of the distribution network, technical requirements and planning and design requirements for microgrid equipment, and requirements of microgrid operation for power quality and reliability. The establishment of microgrid standards should not only ensure security of people and systems, and meet various national rules and regulations, but also allow for noninterference operations; specifically, the integration of DRs should not change the current system and its characteristics, the impacts on the system due to undesired operations should be minimized, including not introducing harmonics, not causing loss of synchronization or scintillation, and not resulting in unintentional islanding. Besides, interconnection rules and standards should apply to all types of DGs and stakeholders.

As the microgrid is directly connected to the distribution network, current standards concerning the distribution network may need to be revised to adapt to the connection of the microgrid. In grid-connected operation, the microgrid may act as a power source or load; therefore, its impacts on the distribution network should be considered. The earthing mode of the microgrid should be coordinated with that of the distribution network. And as the microgrid communicates with the distribution network, the original communication protocols may be changed appropriately due to the impacts of the microgrid. In addition, the active power and reactive power of the microgrid may be independently controlled, which should be considered in specifying power control of the distribution network. The unidirectional energy metering devices in the distribution network should be replaced by bidirectional ones after connection of the microgrid. The provisions on generator protection in DL 400-1991 *Relaying Protection and Security Automatic Equipment* are specific to synchronous motors, which need to be expanded in view of the great variety of DRs and that most DRs are interconnected to the power grid via power electronics. The connection of the microgrid changes the original one-source radial power supply pattern into multisource pattern, making the original line protection no longer suitable and necessitating reconfiguration. The original coordination between the protection of the distribution network and the recloser also becomes unsuitable and needs to be reset. The requirements on interconnection of DRs and on planning and design of the power grid interconnected with DRs set forth in the distribution network standards can be referenced in preparing microgrid standards, with the parameters modified according to the specifics of the microgrid. Microgrid standards need to account for more factors than DR standards as the microgrid is a combination of power sources and loads.

10.4 MICROGRID STANDARD SYSTEM

The standards on the interconnection of DGs to the electric power system are not as mature as those on grid connection of large power stations. As such, the lawmakers, academic institutions, utilities, and power producers need to work together to accelerate the development of microgrid standards. Uniform standards are of great importance to the development of DRs. They will facilitate the commercialization of DR technologies, promote transfer of high-quality products and technologies worldwide, and greatly reduce the costs of traditional energy producers and operators.

The advent of the microgrid changes the traditional radial power supply pattern, and poses challenges to the security and stability of the power grid. Hence, it is imperative to establish systematic microgrid standards to guarantee normalized development of the microgrid.

On November 9, 2009, the expert group CEE-SG was set up, consisting of 20 members from 18 organizations including XJ Group Corporation and Xi'an High Voltage Apparatus Research Institute Co., Ltd. It is responsible for drafting framework standards for smart grid equipment (including microgrid), discussing equipment standardization roadmap, and making action plans.

To increase efforts on the study and establishment of a smart grid standard system (including the microgrid standard system), the SGCC set up a special working group consisting of more than 180 members. Besides, one of SGCC's scientific research programs "Microgrid technology system" also makes studies of the microgrid standard system. After fully considering the basic characteristics of the microgrid, the SGCC proposed a preliminary standard system, comprising equipment specifications, design standards, islanded operation standards, and grid-connected operation standards, as shown in Figure 10.1.

Equipment specifications mainly involve microsources, ESs, inverters, loads (sensitive loads, insensitive loads, and thermal loads), static switches, and protections.

Design standards mainly incorporate provisions on composition, energy management and control of microgrid, types and control modes of microsources, installation position and capacity of ESs, load control, protection configuration, and communication equipment.

Islanded operation standards mainly incorporate provisions on load and generation management, voltage and frequency control, stability, security, protection and control, cold startup, monitoring and communication, power quality, and installation and test.

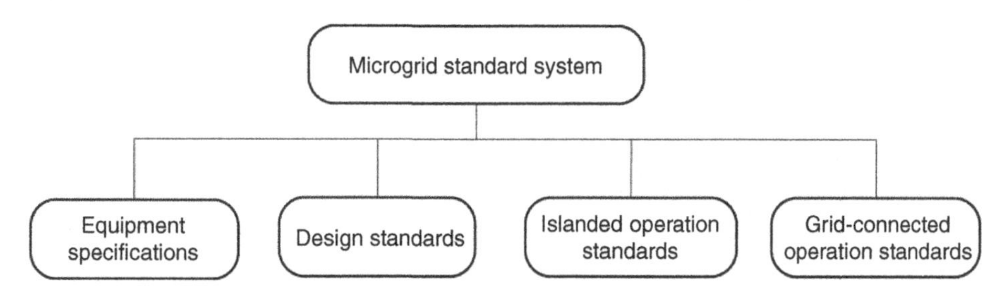

■ **FIGURE 10.1 Microgrid standard system.**

Grid-connected operation standards, directly determining the impacts of connection of the microgrid on the power grid, mainly consist of general principles, power quality, operation and control, security and protection, communication and information, general technical conditions, energy metering, and system test, as detailed here:

1. *General principles*: Provide instructions on the composition, capacity, tie-line power exchange, and energy storage configuration of the microgrid.
2. *Power quality*: Set forth requirements on such indices as harmonics, degree of voltage unbalance, voltage fluctuation and flicker, and injection of direct current.
3. *Operation control*: Set forth requirements for response speed and power change rate in grid-connected operation and islanded operation modes, and during transfer between the two modes.
4. *Security and protection*: Give instructions on configuration of interface protection and internal protection of microgrid.
5. *General technical conditions*: Set forth the technical requirements on electromagnetic compatibility, lightning protection, and grounding.
6. *Energy metering*: Define configuration and metering rules for energy metering devices.
7. *System test*: Specify the position, time, and items of the test on grid-connection of microgrid.

Chapter 11

A practical case

11.1 PROJECT BACKGROUND

In 2010, the State Grid Corporation of China (SGCC) launched the plan for the second batch of pilot programs in its [2010] No. 131 document, and Henan Electric Power Company, as SGCC's only trial distributed generation (DG) connection point, undertook the demonstration program "Comprehensive study on operation control of microgrid containing PV and ES and engineering application" with the 380 kW photovoltaics (PV) power system in the new campus of Henan College of Finance and Taxation.

PV power is of low stability and dispatchability, and may cause difficulty in harmonics management for the grid. To address such problems, the microgrid is set up to allow for distributed generation. Integrating PV power to a distribution network on a local balance basis can facilitate interaction and management of the power grid and loads, and promote utilization of DG, development of smart households, and construction of smart grid and interaction service system. Thus, it is necessary to implement trial projects and conduct studies on the possible maximum integration of DGs to reduce energy consumption and improve energy efficiency, power reliability, disaster prevention capability, and postdisaster emergency supply capability of the entire grid.

11.2 PROJECT DESCRIPTION

The new campus of Henan College of Finance and Taxation is situated in the planned Henan occupational education cluster, 3 km east of the intersection of Beijing–Zhuhai Expressway and Zhengkai Road, Baisha Town, Zhongmu County, Zhengzhou. It has a total floor area of 556,000 m^2 and a planned building area of 342,713 m^2, and over 43,500 m^2 of rooftop areas can be used for PV generation, accommodating a total installed capacity of 2 MW. In view of the design of the seven dormitories, a 380 kW microgrid system combining PV and ES is deployed.

Microgrid Technology and Engineering Application

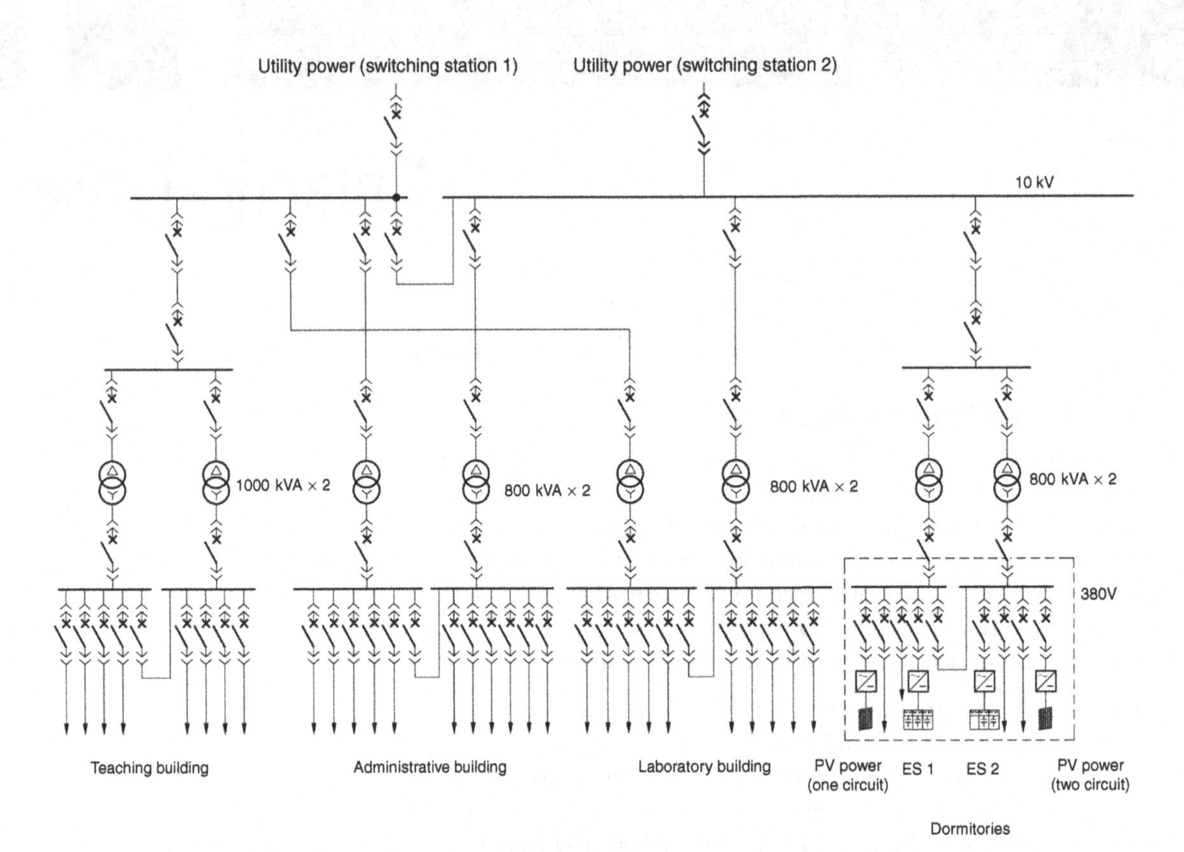

■ **FIGURE 11.1 Schematic diagram of primary connection of the PV–ES combined microgrid system.**

Specifically, the 380 kW PV system is a building-integrated-PV (BIPV) demonstration project jointly launched by the Ministry of Finance and the Ministry of Housing and Urban–Rural Development of China, which will be constructed by Henan College of Finance and Taxation in accordance with given technical requirements; the energy storage (ES) system has a capacity of 2 × 100 kW/100 kWh and uses lithium iron phosphate batteries; the microgrid, including three PV circuits, two ES circuits, and 32 low-voltage (LV) distribution circuits, covers the dormitories and canteens in the No. 4 distribution area of the campus and communicates with the dispatch center of Zhongmu Electric Power Company.

Figure 11.1 shows the primary connection of the microgrid system, where the section circled with a dotted line is the coverage area of the microgrid. In the system, the PV system is connected to the grid on the LV side of the No. 4 distribution area at 380 kV. The DC outputs of PV cells on the rooftop of the seven students' dormitories are first collected and then respectively connected to the two busbar sections on the LV sides of the two distribution transformers in the No. 4 distribution area via inverters.

The two sets of 100 kW/100 kWh lithium iron phosphate batteries are respectively connected to the two busbar sections on the LV sides of the two distribution transformers in the No. 4 distribution area via a power converter system (PCS).

The loads of the system vary seasonally. When the school is open, the power loads may peak at around 600 kW, and the PV power can all be consumed locally; while during the holidays, the loads will fall below 50 kW, when PV generation exceeds power demand, and surplus power will flow to the distribution network.

11.3 SYSTEM DESIGN

11.3.1 Three-layer control system

The control system is of a three-layer architecture, namely, distribution network dispatch layer, centralized control layer, and local control layer, as shown in Figure 11.2. The distribution network dispatch layer, the upper layer, coordinates and dispatches the microgrid to maintain security and economy of the distribution network, and the microgrid is regulated and controlled by the distribution network. The centralized control layer, the middle layer that manages the DRs and various loads in a centralized way, can maintain the microgrid in optimal operation in grid-connected mode

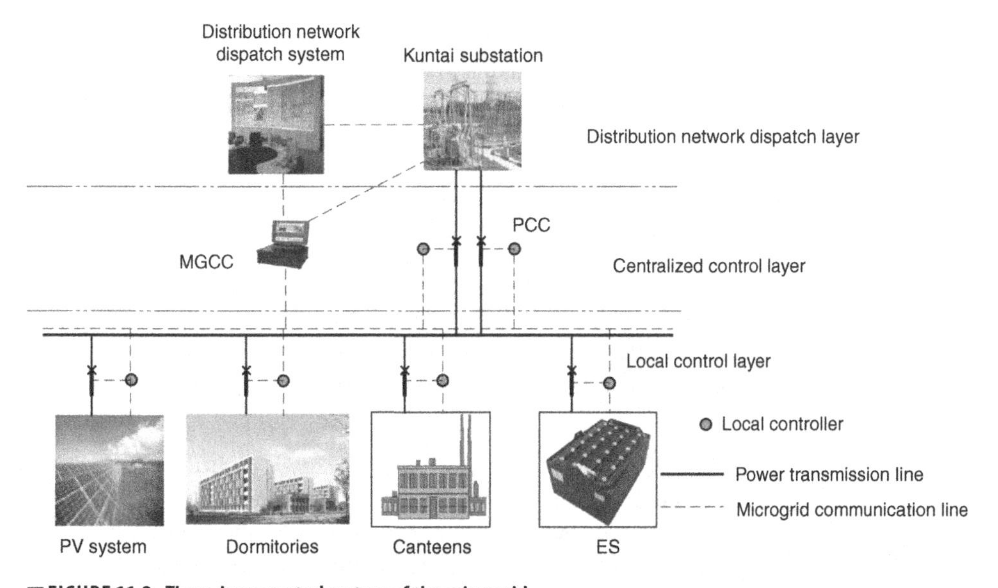

■ **FIGURE 11.2 Three-layer control system of the microgrid.**

and maintain the stability and security of the microgrid in islanded mode by regulating the output of DRs and consumption of various loads. The local control layer, the lower layer, controls various DGs and loads to maintain transient security of the microgrid.

11.3.2 System design

11.3.2.1 PV system

The PV system is connected to the microgrid in three circuits through four inverters, including three 100 kW inverters and one 50 kW inverter. Specifically, two 100 kW inverters are separately connected to the grid, and the other 100 kW inverter and the 50 kW inverter are connected to the grid as one circuit. In addition to their own operation parameters, the inverters are also capable of adjusting their output. The inverters are all installed in the No. 4 distribution area.

The PV power is connected to the grid via an inverter and LV cabinet. The LV cabinet is equipped with electrically operated LV circuit breaker and meters (including power quality meter) to measure the common operation parameters (including voltage, current, active power, and reactive power) and power quality (voltage harmonics and current harmonics of 2nd to 31st orders) of the PV circuits.

11.3.2.2 ES circuits

The two batteries are respectively connected to the two busbar sections via two 100 kW PCSs. In addition to its own operation parameters, the PCS is also capable of output regulation and mode transfer. When the microgrid is out of service, the "black start" function can restore the microgrid back to normal operation quickly.

11.3.2.3 Load circuits

Load circuits are connected to the microgrid through the LV cabinet, which is equipped with an electrically operated LV circuit breaker. Common meters are installed to measure the common operation parameters of the load circuits including voltage, current, active power, and reactive power.

11.3.2.4 PCC circuit

The point of common coupling (PCC) circuit is connected with MSD-831, a grid connection and separation controller that collects the voltage and current of various PCC branches. It can rapidly control the circuit breaker at the PCC, and enables islanding detection, tripping following line faults,

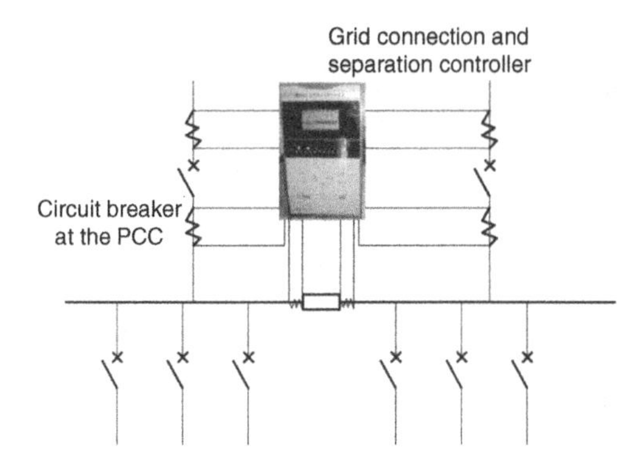

■ FIGURE 11.3 Schematic diagram of the PCC circuit.

synchronous grid-connection after recovery from fault, and autoswitching of standby busbar (see Figure 11.3).

11.3.2.5 LV busbar circuit

The LV busbar circuit is connected with MSD-832, a centralized load controller that collects busbar voltage. At the instant of islanding, it can establish balance between generation and consumption within the microgrid and execute emergency control to remove unimportant loads (or trip some DGs, as the case may be), and during islanded operation, can shed loads at low frequency and low voltage and trip generators at high frequency and voltage to maintain the frequency and voltage of the microgrid within the allowable range (see Figure 11.4).

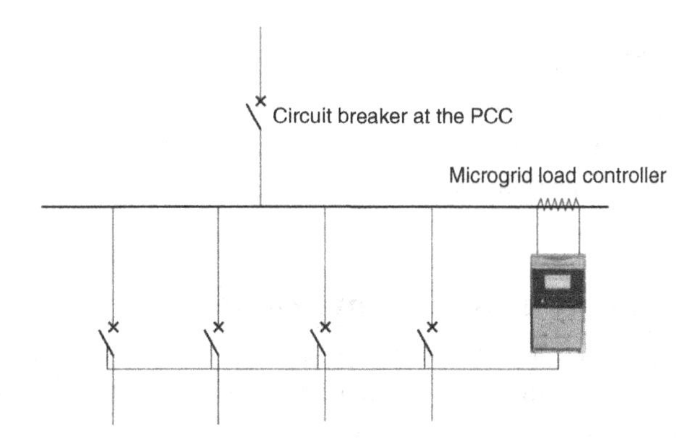

■ FIGURE 11.4 Schematic diagram of microgrid load controller.

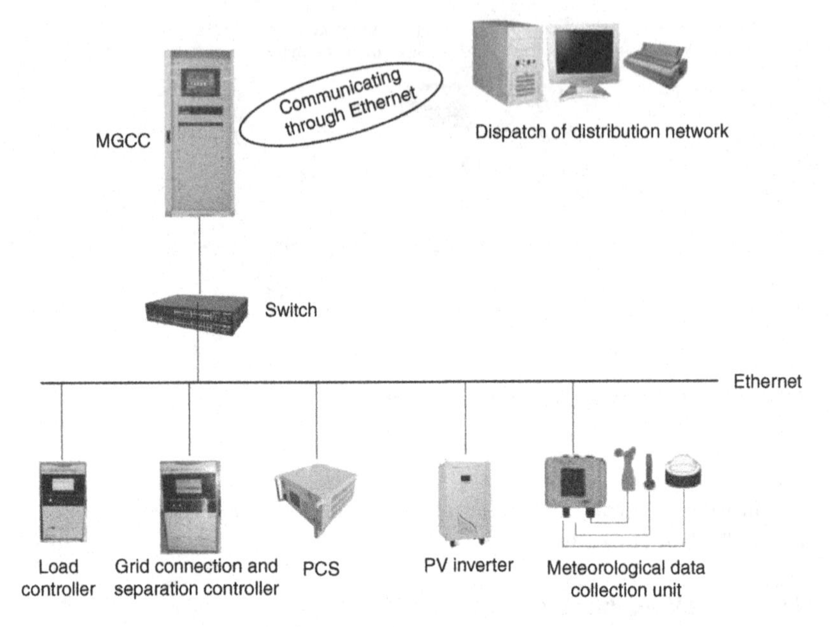

FIGURE 11.5 Structure of the control system.

11.3.2.6 ***Microgrid central controller***

The microgrid control center (MGCC) is equipped with MCC-801 as the centralized controller. Through IEC 61850 communication protocol, it can realize data access, monitoring, and energy management of the entire microgrid. It plays the most important role in microgrid control. As shown in Figure 11.5, MCC-801 communicates with the PV inverter, PCS, grid connection and separation controller, and load controller through Ethernet.

11.3.2.7 ***Monitoring system***

The monitoring system monitors the operation of the entire microgrid by collecting the analog values and Boolean values of the LV measurement and control unit, DG inverter, and grid connection and separation controller in real time.

11.3.3 **Energy management system**

The energy management system makes statistics and advanced analysis of the microgrid in real time on the basis of supervisory control and data acquisition (SCADA). In particular, advanced analysis includes automatic transfer between grid-connected mode and islanded mode, energy dispatch in islanded mode (to automatically maintain balance between generation and consumption), ES charge and discharge curve control, and power exchange control in an emergency (coordinated with the distribution network).

11.3.3.1 Power exchange control in emergency

In special situations (e.g., an earthquake, snowstorm, or flood), the microgrid can serve as reserve to the distribution network, thereby providing effective support to the macrogrid and speeding up recovery of the macrogrid from failure. In this case, the distribution system will specify the amount of power exchange and inform the microgrid, and the MGCC will coordinate the DGs, ESs, and loads according to the specified amount of power exchange.

11.3.3.2 ES charge and discharge curve control

The energy management system defines the expected charge and discharge curves of ESs according to power consumption and PV generation in peak hours and off-peak hours in grid-connected mode, and controls the state of charge (SOC) and charge power and discharge power of ESs in real time according to the curves, thus allowing for load shifting and balance of power consumption and generation.

11.3.3.3 Grid connection and separation control

Upon receiving grid connection orders from the grid connection and separation controller, the energy management system informs the ESs to switch to islanded mode; while upon receiving grid separation orders from the grid connection and separation controller, the energy management system informs the ESs to switch to grid-connected mode.

11.3.3.4 Power balance control in islanded mode

In islanded operation, the energy management system monitors the power generation and consumption of the entire microgrid in real time, restores power supply to loads previously shed, and adjusts the outputs of the PV system and ESs, thus ensuring high reliability and quality of power supply to the most important loads, and keeping the outputs of various DGs and ESs within the allowable range.

To maintain power balance within the microgrid, load shedding or adjustment of DG outputs are required. In load shedding, unimportant loads come before important loads. In adjusting DG outputs, the maximum outputs of renewable energy should be maintained as practical as possible, and then the energy is charged to or discharged from the ESs.

11.3.3.5 Automatic recovery after transfer from islanding to grid connection

When it is detected that the microgrid has been reconnected to the grid, the DGs and loads that were previously removed are reconnected, the DG outputs are adjusted to the maximum, and the ESs are charged, thus restoring the microgrid to normal grid-connected operation and making it ready for possible islanding.

11.3.4 Dispatch of distribution network

Telecontrol equipment is provided at the centralized control layer, which sends the operation information at the PCC to the dispatch automation system of the distribution network. According to economic analysis of the distribution network, the dispatch automation system delivers power exchange regulation orders to the microgrid, making the microgrid a controllable unit of the distribution network.

When the microgrid is in grid-connected operation, the distribution network dispatch layer sends dispatch orders to the microgrid to keep it operating with a specified amount of power exchange, and cooperates with the distribution network in load shifting, economic and optimal dispatch, and rapid recovery from failure.

11.4 OPERATION OF MICROGRID

11.4.1 Operation of the integrated monitoring system

The integrated monitoring system monitors the system voltage and frequency of the microgrid, voltage at the PCC, power of the distribution network, total outputs of DGs, SOC of ESs, and loads within the microgrid. Figure 11.6 shows the operation screens of the integrated monitoring system, Figure 11.7 shows the daily busbar voltage curve of the microgrid, and Figure 11.8 shows the daily system frequency curve of the microgrid.

11.4.2 Operation of the PV system

There are four PV inverters in total, among which the output power of the three 100 kW inverters can be adjusted from 10% to 100%, while that of the 50 kW inverter cannot be adjusted. The startup time of the inverters can be adjusted. To avoid impacts on the PCS due to simultaneous startup of the four inverters in islanded operation, the four inverters are started at different times.

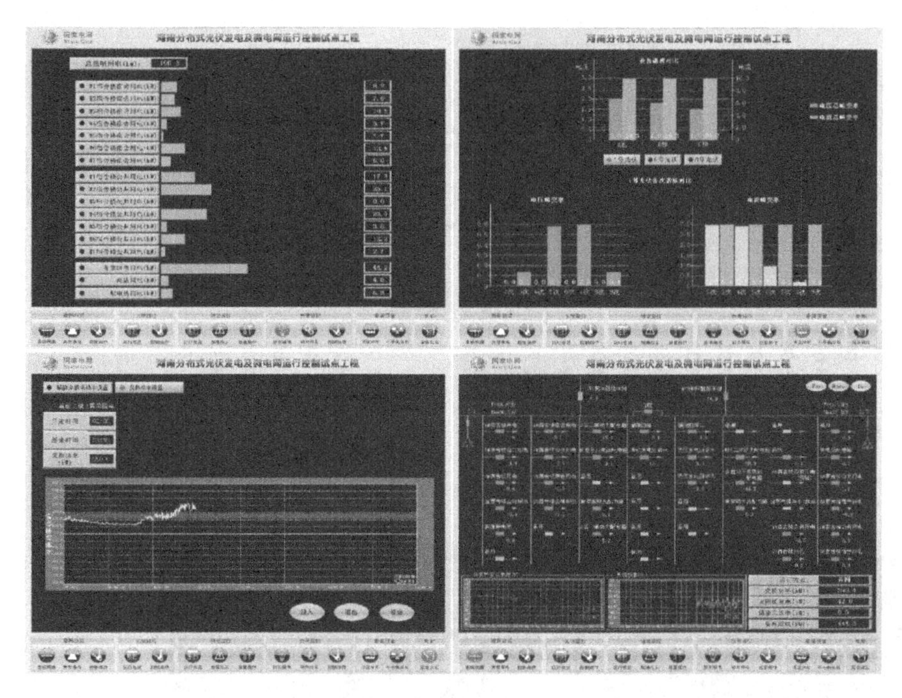

■ **FIGURE 11.6 Operation screens of the integrated monitoring system.**

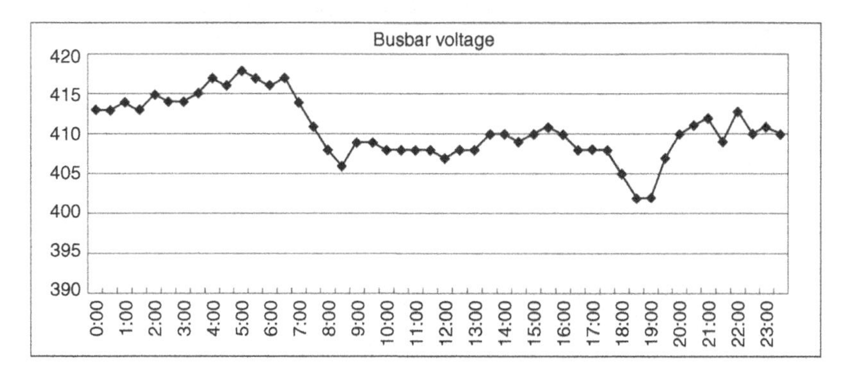

■ **FIGURE 11.7 Daily busbar voltage curve of the microgrid.**

In grid-connected mode, the PV inverters operate at the maximum output; in islanded mode, they operate at the output as specified by the MGCC. Figure 11.9 shows the monitoring screen of operation of the PV system, Figure 11.10 shows the daily output curve of the 100 kW PV inverter, Figure 11.11 shows the voltage waveform of the 380 V busbar in grid-connected mode, and Figure 11.12 shows the voltage spectrum of the

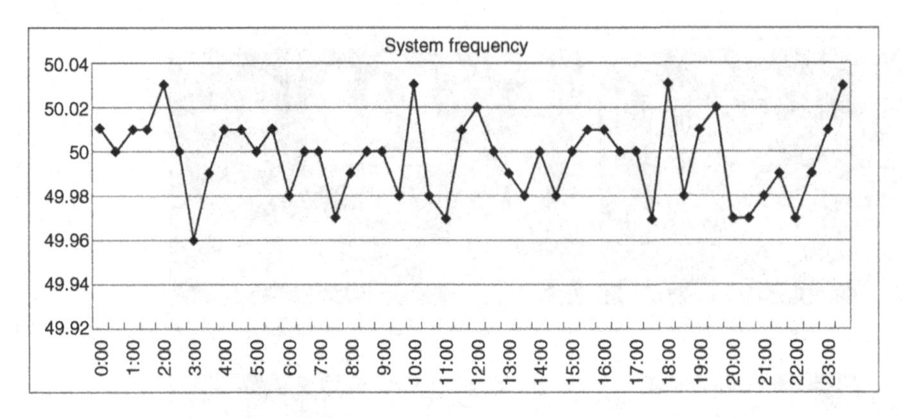

FIGURE 11.8 Daily system frequency curve of the microgrid.

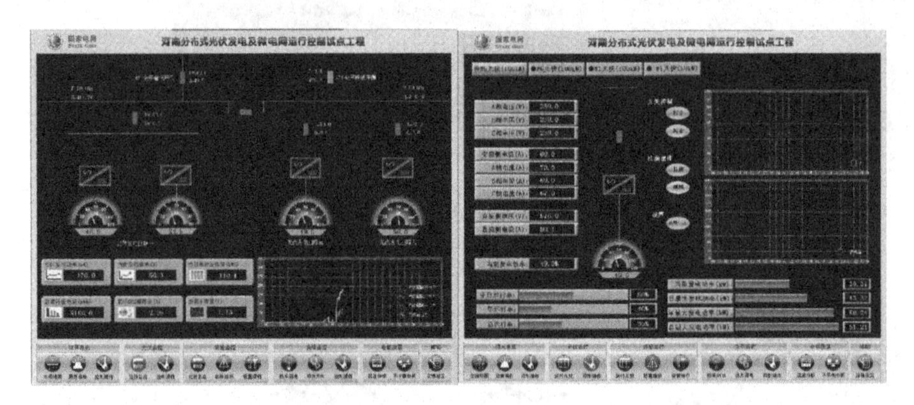

FIGURE 11.9 Monitoring screen of operation of the PV system.

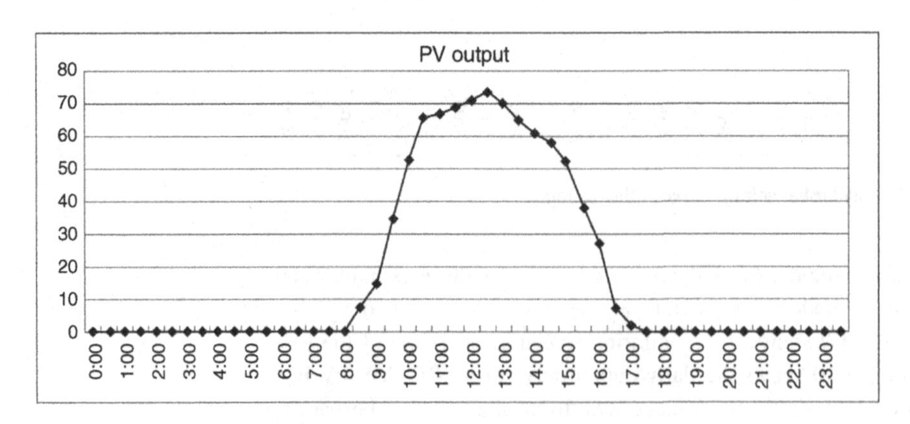

FIGURE 11.10 Daily output curve of 100 kW PV inverter.

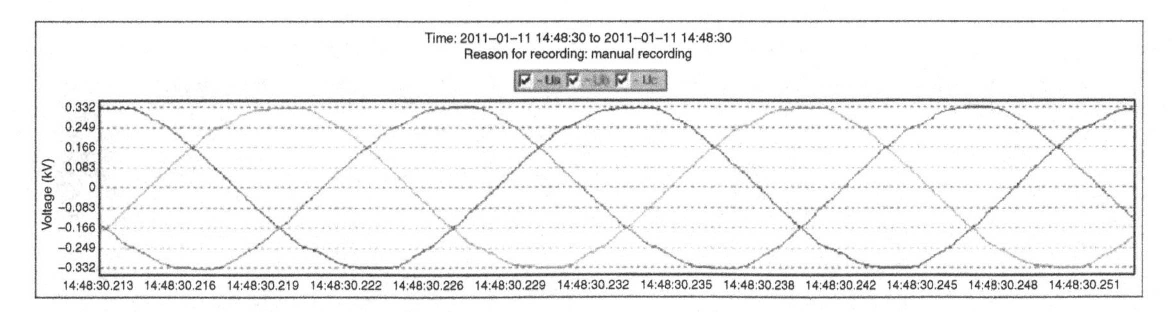

FIGURE 11.11 Voltage waveform of the 380 V busbar in grid-connected mode.

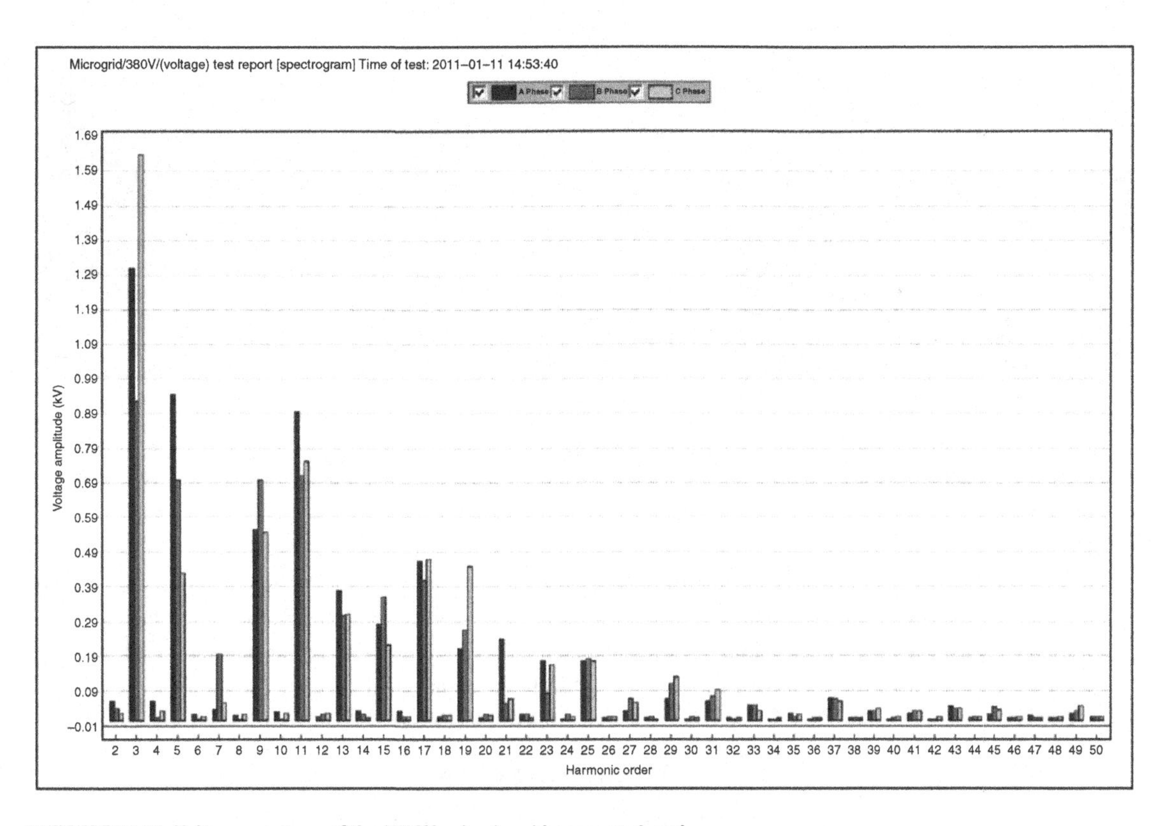

FIGURE 11.12 Voltage spectrum of the 380 V busbar in grid-connected mode.

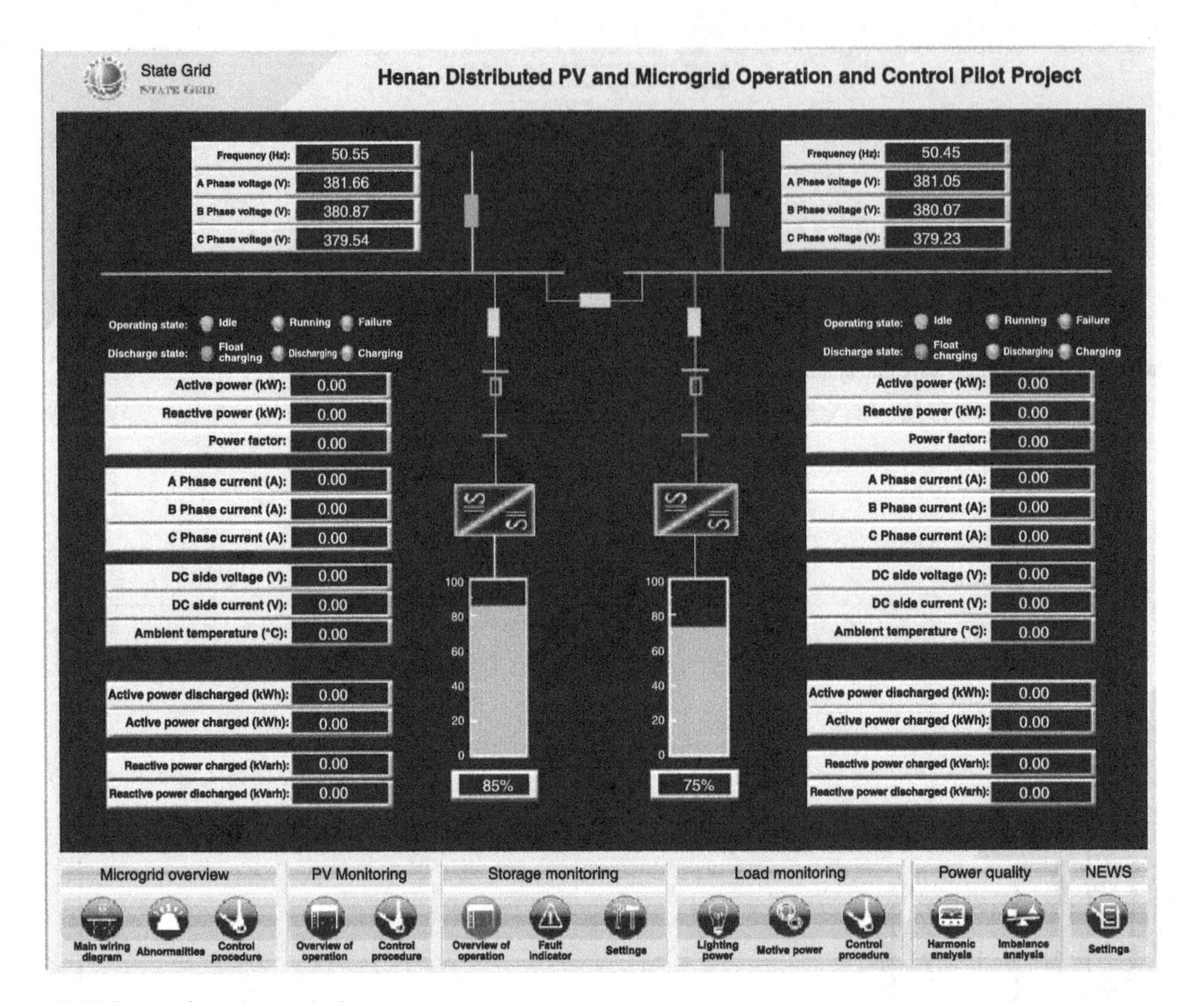

■ FIGURE 11.13 Operation monitoring screen of ES.

380 V busbar in grid-connected mode, where voltage harmonics are mainly odd harmonics of third, fifth, seventh, and ninth orders, and the total harmonic distortion and contents of various orders of harmonics all meet the requirements of applicable national standards.

11.4.3 Monitoring of ES

In grid-connected operation, the PCS operates in *P*/*Q* control and regulates its output as specified by MGCC. In islanded operation, the PCS operates in *U*/*f* control, outputs power at a constant frequency and voltage, and acts as the master power source for the microgrid. Figure 11.13 shows the operation monitoring screen of ES.

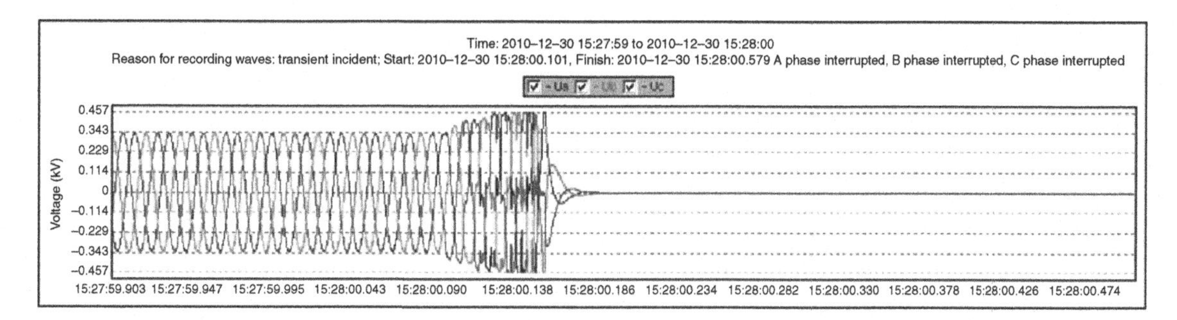

FIGURE 11.14 Voltage waveform of the 380 V busbar during transfer from grid-connected mode to islanded mode.

11.5 TESTS

11.5.1 Test on transfer from grid-connected mode to islanded mode

Purpose: To verify that the microgrid can automatically switch to islanded operation and maintain stability in the mode following an outage in the macrogrid.

Method: Open the switch on the 10 kV side of the distribution transformer to simulate macrogrid outage.

Process: The grid connection and separation controller detects outage of the macrogrid and opens the circuit breaker at the PCC, thus entering the islanded mode. The grid connection and separation controller also sends grid separation signals to the PCS and microgrid control system, and the load control device sheds unimportant loads. After receiving the grid separation signals, the PCS switches to the islanded mode and outputs power at a constant frequency and voltage, and the microgrid control system immediately executes control on the transfer from grid-connected mode to islanded mode.

Figure 11.14 shows the voltage waveform of the 380 V busbar during transfer from grid-connected mode to islanded mode, Figure 11.15 shows the

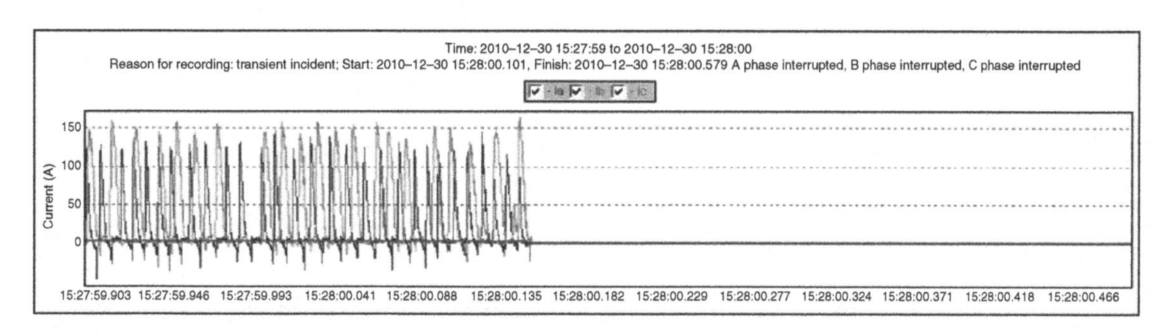

FIGURE 11.15 Current waveform of the PCS during transfer from grid-connected mode to islanded mode.

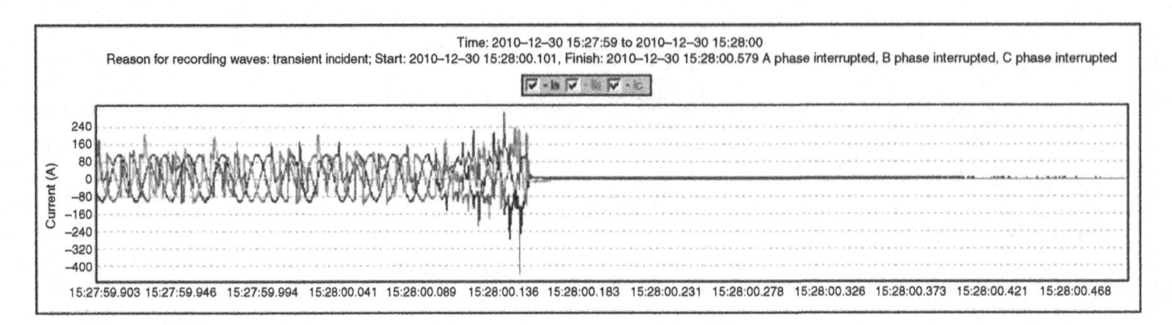

■ FIGURE 11.16 Current waveform of the PV inverter during transfer from grid-connected mode to islanded mode.

current waveform of the PCS during transfer from grid-connected mode to islanded mode, Figure 11.16 shows the current waveform of the PV inverter during transfer from grid-connected mode to islanded mode, and Figure 11.17 shows the voltage trend during transfer from grid-connected mode to islanded mode and then to grid-connected mode. During the transfer from grid-connected mode to islanded mode, due to loss of support from the macrogrid, the circuit breakers of the main incoming line and unimportant

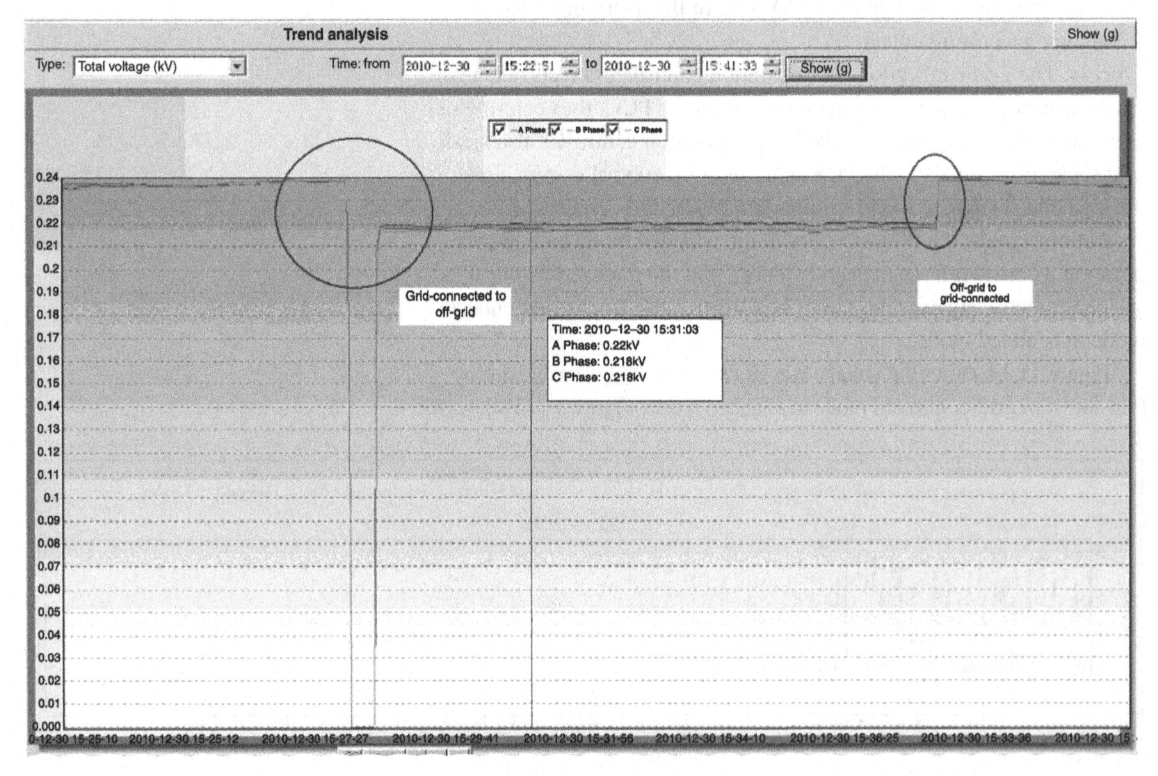

■ FIGURE 11.17 Voltage trend during the transfer from grid-connected mode to islanded mode and then to grid-connected mode.

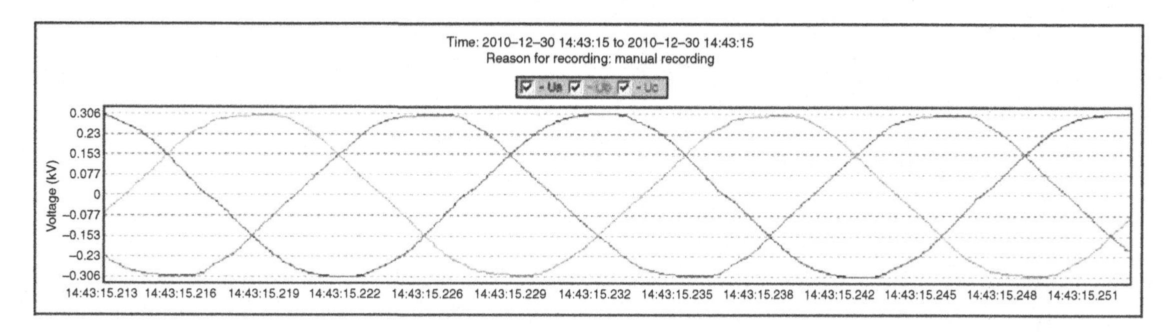

■ FIGURE 11.18 Voltage waveform of the 380 V busbar in islanded mode.

loads open quickly, and instantly the microgrid loses power, the PV system quits service, and the ESs switch from grid-connected mode to islanded mode. Five to 10 seconds later, the microgrid enters the islanded mode, and the busbar voltage and system frequency restore to normal level. As such, after separating from the macrogrid, the microgrid experiences a short-time outage before entering the islanded mode.

11.5.2 Islanded operation test

Purpose: To verify that the microgrid can maintain stability, and achieve optimal output of the PV system and reasonable utilization of ESs by regulating PV power output in islanded mode.

Method: Open the switch on the 10 kV side of the distribution transformer to simulate macrogrid outage, and then the microgrid switches to islanded mode.

Process: In islanded mode, the ESs output power at a constant frequency and voltage to maintain the voltage and frequency of the LV busbar at 380 V and 50 Hz, respectively. The centralized control device of the microgrid regulates output of the PV system, output of auxiliary ESs, and connection and disconnection of loads, to maintain maximum PV output and continuous power supply to important loads, prevent overcharge and overdischarge of batteries, and ensure power supply to less important loads as much as possible. Figure 11.18 shows the voltage waveform of the 380 V busbar in islanded mode, and Figure 11.19 shows the voltage spectrum of the 380 V busbar in islanded mode.

11.5.3 Test on transfer from islanded mode to grid-connected mode

Purpose: To verify that the microgrid can automatically switch to grid-connected mode from islanded mode after the macrogrid has recovered from power failure.

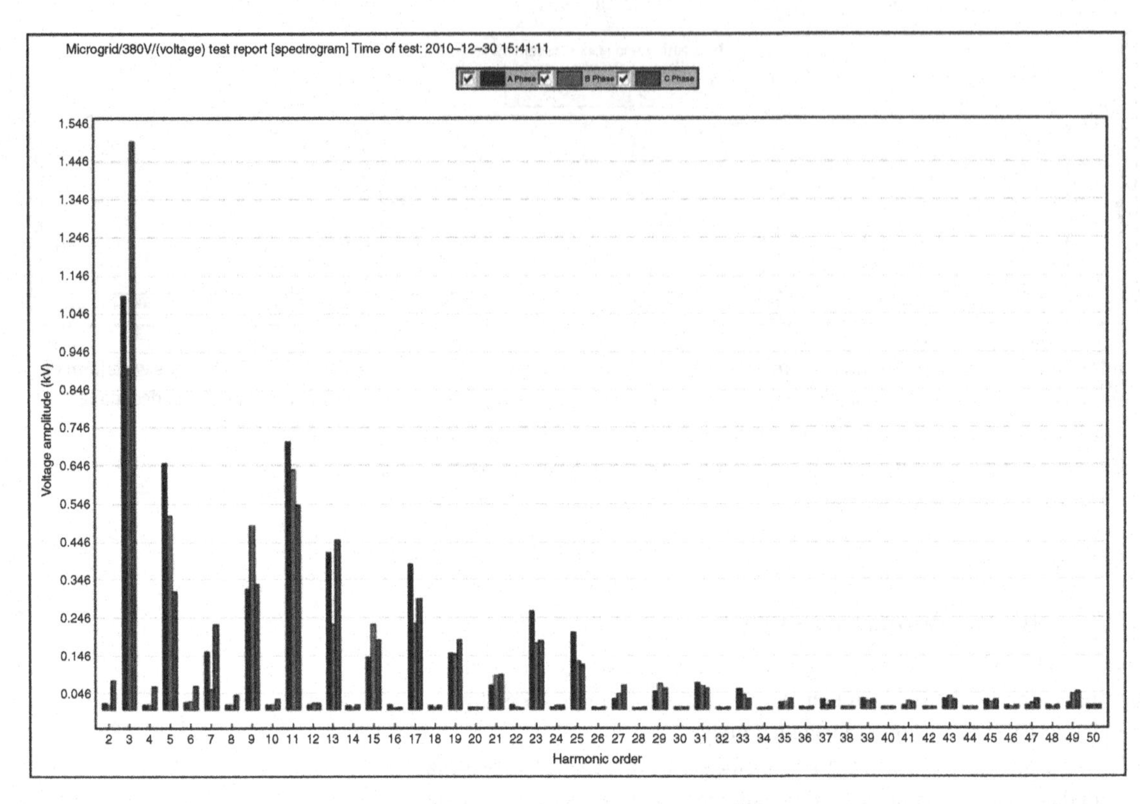

FIGURE 11.19 Voltage spectrum of the 380 V busbar in islanded mode.

Method: Close the switch on the 10 kV side of the distribution transformer to resume power supply by the macrogrid.

Process: When the grid connection and separation controller detects that the voltage on the distribution network side is in the step with the busbar voltage on the LV side in islanded operation, it sends a mode transfer order to the PCS and the PCS switches to *P*/*Q* control from *U*/*f* control, closes the circuit breaker at the PCC, and the control system executes control over the transfer from islanded mode to grid-connected mode. Figure 11.20 shows the voltage waveform on the 380 V busbar during the transfer from islanded mode to grid-connected mode, Figure 11.21 shows the current waveform of the PCS during transfer from islanded mode to grid-connected mode, and Figure 11.22 shows the current waveform of the PV inverter during the transfer from islanded mode to grid-connected mode. When the microgrid transfers from islanded mode to grid-connected mode, the voltage of the 380 V busbar drops and instantly resumes normal level, the PV inverter realizes low-voltage ridethrough, thus ensuring uninterrupted PV generation, and the ESs also resume normal operation, thus realizing smooth transfer from islanded mode to grid-connected

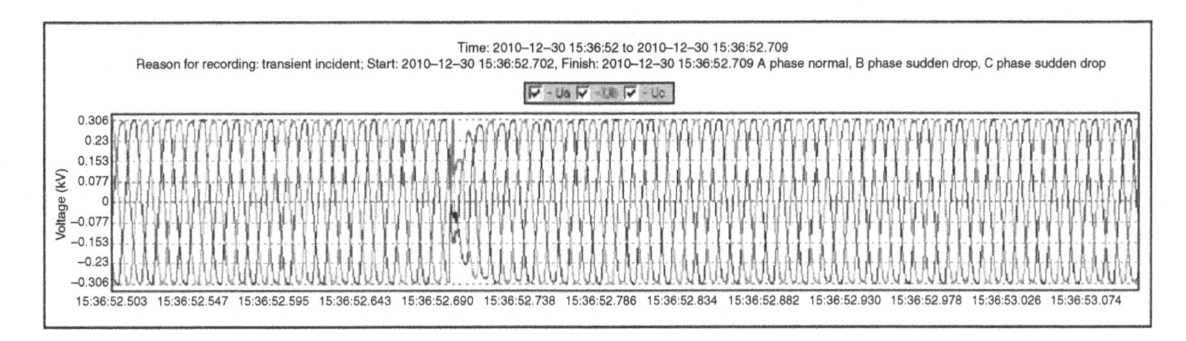

■ FIGURE 11.20 Voltage waveform of the 380 V busbar during transfer from islanded mode to grid-connected mode.

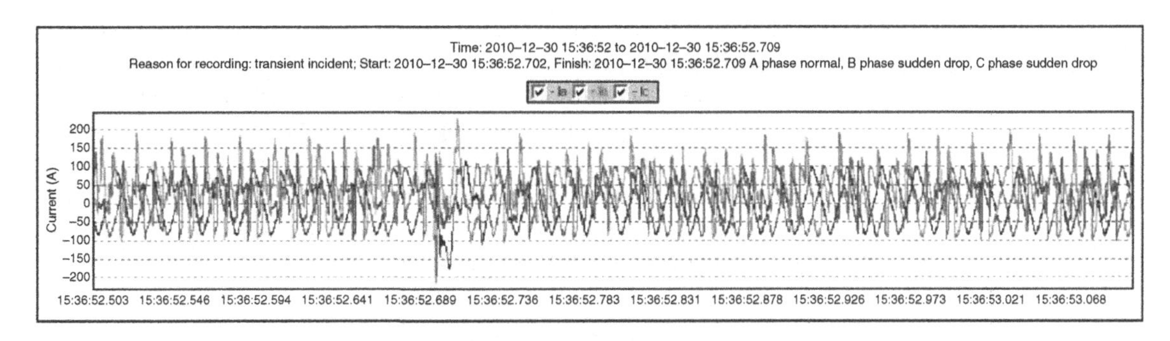

■ FIGURE 11.21 Current waveform of the PCS during transfer from islanded mode to grid-connected mode.

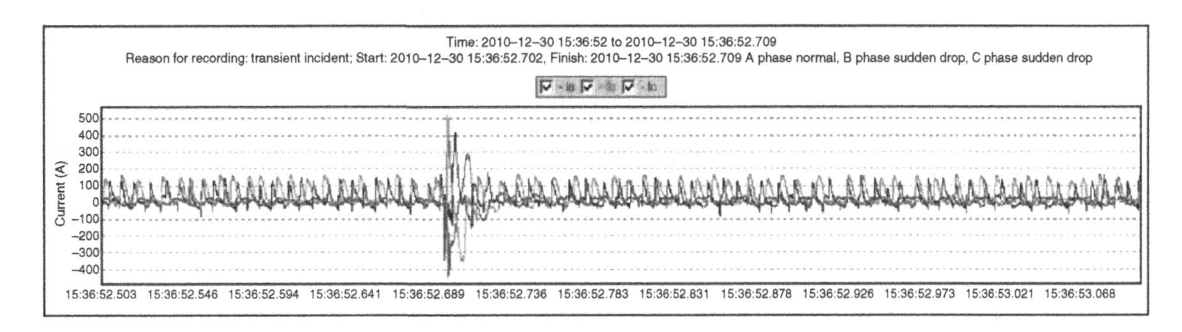

■ FIGURE 11.22 Current waveform of the PV inverter during transfer from islanded mode to grid-connected mode.

mode. The voltage trend during the transfer from islanded mode to grid-connected mode is shown in Figure 11.17.

11.5.4 Grid-connection recovery test

Purpose: To verify that the microgrid can operate normally after being reconnected to the grid.
Process: When the microgrid is reconnected to the grid, the centralized control device of the microgrid reconnects unimportant loads previously shed in islanded mode, regulates the output of the PV inverter to the maximum, and automatically charges the ES to prepare for possible islanding.

11.5.5 ES charge and discharge control

Purpose: To verify that the ESs can be charged or discharged as required. Specifically, in peak hours, the ES can automatically discharge energy and in off-peak hours, the ES can automatically charge, thus realizing load shifting for the grid and improving the operation environment of the grid.
Process: Set the charge power and discharge power of the ESs according to peak hours and off-peak hours of the system to establish the charge and discharge power curve, and the control system automatically controls the output of the ES in real time according to the curve.

11.5.6 Power exchange control

Purpose: To test the response of the microgrid system to power exchange dispatch orders from the upper level.
Method: Manually set the value of power exchange to simulate dispatch orders delivered from the upper level.
Process: Set the target output power, and the system automatically regulates the PV output and ES, or even temporarily sheds unimportant loads (in emergency cases), to keep the actual power exchange as close as possible to the specified value, and the microgrid can provide support for the distribution network.

Appendix: List of abbreviations

ASON	Automatic switched optical network
BMS	Battery management system
CAES	Compressed air energy storage
CCHP	Combined cooling, heat and power
CERTS	Consortium for Electric Reliability Technology Solutions
CHP	Combined heat and power
DER	Distributed energy resources
DG	Distributed generation
DR	Distributed resource
EPON	Ethernet passive optical network
EPS	Emergency power system
ES	Energy storage
FACTS	Flexible AC transmission systems
FRIENDS	Flexible reliability and intelligent electrical energy delivery system
GOOSE	Generic object oriented substation event
IEC	International Electrotechnical Commission
IED	Intelligent electronic device
IEEE	Institute of Electrical and Electronics Engineers
MG	Microgrid
MGCC	Microgrid control center
MMS	Manufacturing message specification
MPPT	Maximum power point tracking
MSTP	Multiservice transport platform
OAM	Operation administration and maintenance
OTN	Optical transport network
PCC	Point of common coupling
PCS	Power converter system
PE	Protecting earthing
PLC	Power line communication
PTN	Packet transport network
QoS	Quality of service
RCD	Residual current device
SDH	Synchronous digital hierarchy
SMES	Superconducting magnetic energy storage
SNTP	Simple network time protocol
SOC	State of charge
SV	Sampled value
TCM	Tandem connection monitoring
TDM	Time division multiplex
THD	Total harmonic distortion
UPS	Uninterrupted power supply
VPP	Virtual power plant
VSI	Voltage source inverter
WDM	Wavelength division multiplexing

References

[1] Zhenya L. Smart grid technology. Beijing: China Electric Power Press; 2010.
[2] Li R, Zhou F, Li Y. Ground PV power system and its application. Beijing: China Electric Power Press; 2011.
[3] Li G. Power system transient analysis. Beijing: China Electric Power Press; 1985.
[4] Wang Z, Yang J, Liu J. Harmonic suppression and reactive power compensation. Beijing: China Machine Press; 2005.
[5] Dai X. DC transmission base. Beijing: China Water Power Press; 1990.
[6] Wu J, Sun S. Harmonics in power system. Beijing: China Water Power Press; 1998.
[7] Jiang Q, Xie X, Chen J. Parallel compensation of electric power system – structure, principle, control and application. Beijing: China Machine Press; 2004.
[8] Zhao W. HVDC engineering technology. Beijing: China Electric Power Press; 2004.
[9] DC Transmission Research Group of Zhejiang University. DC transmission. Beijing: China Electric Power Press; 1982.
[10] Pocce AB. HVDC system structure and operating modes. Translated by HVDC Research Center of North China Electric Power University. Beijing: Water Resources and Electric Power Press; 1979 (In Russian).
[11] Arrillag J. HVDC Transmission. Translated by Ren Zhen, et.al. Chongqing: Chongqing University Press; 1987 (In English).
[12] Mao J, Zhou F, Ma H. Optimal design of microgrid networking. North China Elec Power 2012;1:32–5.
[13] Zhang Y, Li Q, Li Z, Yang H, Ma H, Li M. Conceptual design for the project of microgrid system interconnected with photovoltaic generation & energy storage. Power Syst Protect Control 2010;38(23):212–4.
[14] Zhang Y, Li X. Research on microgrid centralized control strategy based on active power vacancy. Power Syst Protect Control 2011;39(23):106–11.
[15] Shu H. Study on operation characteristics and control of microgrid (a master's thesis). Shanghai: Shanghai University of Electric Power; 2011.
[16] Yaodan. Research on islanding of distributed generation system (a master's thesis). Hefei: Hefei University of Technology; 2006.
[17] Du X, Huang Qi, Zhang C, Jing S, Zhou D. Overview of grid-connected inverting techniques based on microgrid. Zhejiang Elec Power 2009;4:17–21.
[18] Chen C, Mao C, Wang D, Lu J. Park transformation of multiphase AC systems. High Voltage Eng 2008;34(11):2475–81.
[19] Zhang X. Study on novel current control strategy for voltage source PWM rectifier with LCL filter. Large Power Conversion Technol 2009;6:36–40.
[20] Ju H. Research on the parallel control of multi-inverters in distributed micro-grid power system (a master's thesis). Hefei: Hefei University of Technology; 2006.
[21] Niu C. Microgrid islanding detection and islanding division (a master's thesis). Tianjian: Tianjin University; 2008.
[22] Wu T, Mao X, Xiang J. Microgrid solar power system islanding detection. J Hubei Univ Technol 2010;25(1):19–21.

Microgrid Technology and Engineering Application.

[23] Ding L, Pan Z, Su Y, Cong W. Splitting and islanding of networked dispersed generators. Elec Power Automat Equip 2007;27(7):25–9.
[24] Xue Y, Tai N, Liu L, Yang X, Jin N, Xiong N. Coordination control strategies for islanded microgrid. China Elec Power 2009;42(7):36–40.
[25] Yang X, Su J, Ding M, Du Y. Research on frequency control for microgrid in islanded operation. Power Syst Technol 2010;34(1):164–8.
[26] Lin X, Li Z, Bo Z, Andrew K. An under-frequency load shedding method adaptive to the islanding operation of microgrid. Power Syst Technol 2010;34(3):16–20.
[27] Li D, Chen Q, Jia Z. A practical series hybrid active power filter based on fundamental magnetic flux compensation. Trans China Electrotechnical Soc 2003;18(1):68–71.
[28] Duan Y, Wang Y, Fu Z, Yang J, Wang Z. Study on a novel parallel hybrid power filter. Adv Technol Elec Eng Energy 2004;23(1):51–4.
[29] Yang Z, Wu C, Wang H. Design of three-phase inverter system with double mode of grid-connection and stand-alone. Power Electr 2010;44(1):14–6.
[30] Wei W, Xu S. Research of grid-connected photovoltaic inverter. Power Electr 2008;42(11):43–4.
[31] Zhou Y, et al. Digitizing design of 3-phase SPWM bi-directional inverter voltage control. Power Electr 2004;38(1):42–4.
[32] Ding M, Yang W, Zhang Y, Bi R, Xu N. IEC 61970 based microgrid energy management system. Elec Power Automat Equip 2009;29(10):16–20.
[33] Ding M, Wang M. Distributed generation technology. Elec Power Automat Equip 2004;24(7):31–6.
[34] Lu Z, Wang C, Min Y, et al. Overview on microgrid research. Automat Elec Power Syst 2007;31(19):100–7.
[35] Wang J, Li Z, et al. Technical standards and related issues of microgrid. East China Elec Power 2011;39:1612–4.
[36] Xue Y, Tai N. Introduction to existing DG interconnection standards worldwide. Southern Power Syst Technol 2008;1.9:13–7.
[37] Han Y, Zhang D, et al. A study on microgrid standard system in China. Automat Elec Power Syst 2010;34(1):69–71.
[38] Walmir F, Ze H, Xu W. A practical method for assessing the effectiveness of vector surge relays for distributed generation applications. IEEE Trans Power Syst 2005;20(1):57–63.
[39] Kai S, Da-Zhong Z, Qiang L. A simulation study of OBDD-based proper splitting strategies for power systems under consideration of transient stability. IEEE Trans Power Syst 2005;20(1):389–99.
[40] Pecas Lopes JA, Moreira CL, Madureira AG. Defining control strategies for micro grids islanded operation. IEEE Trans Power Syst 2006;21(2):916–24.
[41] Hatziargyriou N, Asand H, Iravani R, et al. Microgrids. IEEE Power Energy Mag 2007;5(4):78–94.

Index

D

E

F

G

Printed in the United States
By Bookmasters